CITADELS
RAMPARTS
&
STOCKADES

Also by Irvin Haas

America's Historic Houses and Restorations
America's Historic Inns and Taverns
America's Historic Villages and Restorations
America's Historic Ships, Replicas and Restorations
Historic Homes of the American Presidents
A Treasury of Great Prints

CITADELS
RAMPARTS
&
STOCKADES

America's Historic Forts

by

Irvin Haas

EVEREST HOUSE
Publishers   *New York*

for my wife
Irene

Copyright © 1979 by Irvin Haas
All Rights Reserved
Library of Congress Catalog Card Number: 78-74583
ISBN: 0-89696-038-2
Published simultaneously in Canada by
Beaverbooks, Pickering, Ontario
Manufactured in the United States of America by
Halliday Lithograph Corporation
West Hanover, Massachusetts
Designed by Sam Gantt and Steve Peterson
First Edition

CONTENTS
Middle Atlantic

New England

Midwest and West

South

Preface

FROM THE CRUDE AND PATHETIC little fort built by a handful of colonists from Elizabethan England in the wilderness of Virginia in 1585 to the massive fortress-towns erected by the government or the large and affluent trading companies on our western frontiers and coastlines, the history of the fort is the story of the settling and expansion of America.

The history of the fort in America started when some European nations became aware of the rich benefits of explorations in the New World. In the middle of the sixteenth century they claimed land in North America by settlement. France and Spain started to colonize the southeastern Atlantic seacoast. France concentrated on the St. Lawrence River and England on the central coastal region. Colonizers were a precious asset, and for their protection from Indians and other Europeans, fort construction was a necessary defensive measure. They located the forts strategically on harbors, on rough interior highways, at the mouths of rivers, and in most early land claims. The fort enabled them to control large areas. Without these forts our early pioneers would have remained huddled on some narrow shore hugging the protective water, or at the edge of some forest, ready in time of peril to hide, retreat, or simply dig in. The fort was their blessed sanctuary, but it was also their secure staging area to advance, explore, or fight.

The fort in America took many shapes and designs. They were constructed of whatever indigenous materials were available. The early forts were collective efforts built by collective initiative and of expedient construction because there was little money, manpower, or engineering skill. But whatever forms they took and whatever the designs and construction, they enabled our forefathers to come to practical terms with their situations and environments. These were the structures they devised to protect themselves, and, in many cases, they settled down to a permanent life behind its sheltering walls. In addition to the all-important trading post, there were permanent domiciles in many of the trading forts. They supported a population devoted to serving the sponsoring company. The trading fort was both a home and a base to protect commercial expansion.

The military forts were more transient things. They either led or followed the pioneers across the continent. They were not designed to resist siege but rather to serve as bases from which Indians could be pursued and punished for attacks on soldiers, settlers, and others. Their mere existence was a deterrent to Indian attacks. They also served as watering places and wagon stops for the numerous wagon trains trekking west by way of the new roads opened with the aid of the military and their forts. These forts also brought commerce to the West providing bases where sutlers could provide supplies and refreshments. Some of the forts became stations for the Pony Express or stagecoach stops for tired travelers. But all of them served as grim reminders of the might and power of the white man's government in Indian country and served as rough centers and jails for frontier justice.

A little-known aspect of the forts' influence

on its place and time is that they aided in the development of Western culture. They maintained libraries and reading rooms, and some of them had schools. Post chapels provided facilities for religious services and weddings. Post bands provided entertainment. Some of the forts had gardens to reduce expenses and were among the first pioneers in American experimental agriculture.

The media has glamorized the fort in countless fairy-tale movies and in thousands of Western novels. Each had a kernel or a bit more than a kernel of truth and reality. History cannot diminish the glorious part they played in the saga of the United States. They withstood the Indian attacks, the pirate attacks, the attacks of the Spanish, French, and British, the attacks of the Blue and the Gray, and it is to the few proud survivors that we dedicate this book. The fort was a very transient thing; abandoned when peace came to their particular territory, they became victims to decay and vandalism.

They were destroyed and ravaged for their planks of timber and pieces of stone. Others were put to civilian use and altered beyond recognition. Fortunately, like so many other historic sites today, the remaining forts are receiving the protective sponsorship of local historical societies, state agencies, or the National Park Service. They are being preserved, reconstructed, or duplicated so that they might continue to stand as symbols of the pioneering spirit that built the United States.

I acknowledge with gratitude the always understanding and generous cooperation of the National Park Service who administers an impressive group of our national landmarks. I want to thank the various local historical and military societies who supplied data on their sites and who are preserving and maintaining forts in their respective communities that have played important and vital roles in their local histories. I trust that in some way this book will serve to laud and perpetuate their efforts.

Photo Credits

Division of Beaches and Parks, State of California; Arkansas Department of Parks and Tourism; Russ Hanson; Robert T. Hotelling; Historical Society of Delaware; Fort Ticonderoga Museum; Kansas Historical Society; Kansas City Historical Society; Paul Lopes; Fred W. Marvel; Michigan Tourist Council; Microfilm Division, Department of State, Georgia; Minnesota Department of Economic Development; Grace McClellan; Clay Nolen; New Mexico Department of Development; New Jersey Department of Conservation and Economic Development; National Park Service; Ohio Historical Society; Office of Vacation Travel, New Hampshire; State Historical Society, North Dakota; State Historical Society, Colorado; U.S. Army; Delmott Wilson; W. Emerson Wilson; Plimoth Plantation; Wyoming Travel Commission; Wyoming State Archives.

Middle Atlantic

Fort Augusta

SUNBURY, PENNSYLVANIA

Throughout the mid-eighteenth century, the British and the French were engaged in a struggle for control of western Pennsylvania. The junction of the West and North Branches of the Susquehanna River in present Northumberland County provided a strategically valuable position for the British, and it was natural that they should construct a fortification there. This fort, presumably named after Princess Augusta, then the widow of the Prince of Wales and the mother of the future George III, was the largest fort built by Pennsylvania during the French and Indian War, the one longest garrisoned by Provincial troops, and the last one used for military purposes.

Fort Augusta was built in answer to the need to contain the French and their Indian allies after their resounding victory over General Edward Braddock in July 1755. The massacres on Penns Creek near present Selinsgrove the following October further intensified the danger in the Susquehanna Valley. Construction of the fort was begun in July 1756 under Colonel William Clapham and was completed by Major James Burd the following year.

The fort formed a square with bastions at the four corners. It was 204 feet in length including the bastions. Constructed of logs, it included a moat, an outer stockade, and four blockhouses connected in such a way as to form a covered pathway to the river.

Due to its physical size and the size of the garrison, Fort Augusta was never in any danger of enemy occupation. A regiment consisting of eight companies of fifty men each was almost always present, although the number of men was not constant. Occasionally groups were sent on forays in the vicinity, but the garrison did not actively seek the enemy. The fort became a center of peaceful Indian activity, providing a base for sending representatives to confer with the Indians, and also as a stopping point for the Indians themselves on their journeys to treaties and conferences. Trade with the Indians was encouraged by the establishment of a trading post at the fort.

In 1758 a reorganization of Pennsylvania's defenses by the British tied Fort Augusta more closely to the forts built earlier, and men from its garrison accompanied Forbes on his expedition to Fort Pitt in 1758. After 1761 Fort Augusta was the only frontier fort maintained by Pennsylvania, and when the new hostilities of Pontiac's War broke out in 1763, it was the only Provincial garrison prepared to meet the renewed threat of Indian attack. During the American Revolution it also served as a base for General Sullivan's expedition against the Iroquois in 1779.

Fort Augusta was abandoned after the Revolution, and gradually fell into ruins, except for the commandant's quarters. These were occupied by Colonel Samuel Hunter, the last commander of the fort. In 1852 the commandant's headquarters burned down, and a grandson of Colonel Hunter built the house that presently stands behind the model of the fort.

The fort model on the site, reproduced at an approximate scale of 1:6, includes all features of the fort except the blockhouses and the covered way. The original well and powder magazine may also be seen at the present site. The Hunter Mansion now serves as a museum, in which are displayed cannon and other relics of this frontier fort of the Susquehanna Valley.

You can reach Fort Augusta from Williamsport by taking U.S. 15 south to Sunbury and

then 147 to the fort. You may go north from
Sunbury on 147 to the fort. It is open from
May to September, Tuesday to Saturday, 9:00
A.M. to 5:00 P.M., Sunday 1:00 P.M. to 5:00 P.M.;

from October to May, Tuesday to Saturday
10:00 A.M. to 4:30 P.M., Sunday 1:00 P.M. to
4:30 P.M. No admission fee.

Fort Delaware

NARROWSBURG, NEW YORK

FORT DELAWARE is a replica of the original stockaded settlement of Cushetunk.
The original fort was built about 1755 and was
active until about 1785. The original site is
about six miles north of its present location.
Because the original site is frequently flooded,
it was decided to place the replica in Narrowsburg.

The fort was built to protect the newly established farmers who came from Connecticut to
buy land in the fertile Delaware Valley.
Visitors to Fort Delaware usually ask a single
question after viewing the site: "Why should
good, God-fearing, industrious people of the
Colony of Connecticut want to leave their com-
fortable homes in well-settled communities and
take up lands in Indian country, some forty
miles west of the recognized boundary between
settler and Indian?" The reason seems rather
hard to believe, but nevertheless true; by 1750
the Colony of Connecticut became overpopu-
lated, especially in the extreme eastern part of
the state. There was no longer any farmland
available in that region for young families to
acquire. The large families of the early colonists contributed largely to the land shortage.
The original grants were about 600 acres. If the
grantee was blessed with four sons and several
daughters, and this size was rather common,
the original grantee would arrange to divide the
original acreage between his sons, and they
would become eligible for 150 acres. In due
time the grandchildren would gain title to lesser

plots and so on. The resultant small farms
provided a bare livelihood. The farmers of 1750
had no knowledge of contour farming, crop
rotation, chemical fertilizers, etc. When their
small farms failed to yield, it was time to pick
up stakes for larger parcels of land. Hence their
trek to the Susquehanna and Delaware valleys.

The Delaware Company, a private land concern, purchased a vast area of land from the
Lenape Indians, and Cushetunk is the name
that was formerly applied to that region. The
principal village of the region is now known as
Cochecton. In 1760 the proprietors of the Delaware Company had erected three towns, each
extending ten miles along the Delaware River.
The Cushetunk Settlement had thirty cabins,
three log houses, a gristmill, and a sawmill.
The settlers had never been quite free from
attacks by the Indians. In 1763 the Delawares,
in reprisal for the burning to death of their
great Sachem in his wigwam, organized war
parties and destroyed most of the homes along
Ten Mile River and murdered all of the inhabitants.

During the Revolutionary War Cushetunk
was an exposed and isolated region. The Mohawk and Seneca Indians were allied with the
British. Their small war parties harassed the
settlers. In 1778 Indians, Tories, and British
Rangers occupied the Cushetunk fortifications.
After the Revolution the dwellings and the old
stockade fell into disuse and decay, and it is not

Moat and drawbridge of Fort Delaware (Photo: W. Emerson Wilson)

Inside walls of Fort Delaware (Photo: Henry J. Szymanski)

recorded when they disappeared from the scene.

Since 1970 Fort Delaware has been operated as a municipal museum. Built to the specifications described in extant records of the Cushetunk Settlement, the Fort consists of a stockade, surrounding dwelling cabins, blockhouses, gun platform, storehouses, blacksmith shed, candle shed, armory, and animal yard. Within the stockade proper are three cabins which represent the dwellings of the early settlers. Each cabin depicts a different type of dwelling. The most primitive shows the dwelling of the lone trapper and hunter. It is built as the original would have been with bark roofing, fireplace, and few artifacts. The other two cabins are more sturdy and show the influence of women and children in the fort. Each cabin displays household and other living necessities. Each of the blockhouses are equipped with articles pertaining to styles of colonial life. There is a collection of early muskets, and the staff gives demonstrations of firing the muskets and also the firing of a muzzle-loading cannon.

There is an herb and vegetable garden and animals in the stocks that you would have found in this area in 1755. Staff members provide lecture-demonstrations illustrating many of the day-to-day tasks performed by the first colonial settlers.

Fort Delaware can be reached on Route 97 to Narrowsburg, New York. It is open daily from 10:00 A.M. to 5:30 P.M. from the last Saturday in June through Labor Day. Adults $1.50. Children 75¢.

Fort Le Boeuf

WATERFORD, PENNSYLVANIA

THE HISTORY OF FORT LE BOEUF dramatically reflects the struggle between France and Britain for colonial empire, as well as the later triumph of an independent America. The site's strategic location made it of vital importance in the struggle for control of the wilderness in the early eighteenth century. As the French were later displaced by the British and the British in turn by the Americans, the position along French Creek became the site of four successive forts.

Control of the Ohio Valley, including present western Pennsylvania, was essential to the French in linking their domain with Canada and with that in Louisiana. When the English traders began to expand their operations into present Ohio, the French determined to erect a series of forts southward from Canada to assure control in this area of conflicting claims. In 1753 the French built Fort de la Rivière aux Boeufs at the end of an early Indian portage path to the Rivière au Boeuf (French Creek), which emptied into the Allegheny River.

The French fort was square, with bastions at each corner. Barracks and other buildings formed the sides of the square; built into the bastions were a guardhouse, chapel, infirmary, and the commander's storehouse. The outer stockade was constructed of vertical wooden posts. Illness and the difficulty of transporting men and supplies delayed the completion of the fort, so that the French were not able to continue down the river to erect a fort at the Forks of the Ohio (Pittsburgh) as quickly as had been their plan.

The increased activity of the French alarmed the English. Governor Robert Dinwiddie of Virginia sent a personal message to the commander at Le Boeuf stating British claims to the area and demanding peaceful withdrawal of French forces. The young Virginia messenger who volunteered to deliver this message was George Washington, who at the age of twenty-one received a harsh initiation into wilderness travel. Although the French refused to evacuate, the journey gave Washington an opportunity to observe both the fort and French preparations for traveling down the Ohio. Washington's report to the Virginia government alerted the British to the situation, which grew rapidly into the French and Indian War.

From 1755 to 1758, Fort Le Boeuf served as a way station on the French line of defense to Fort Duquesne (Pittsburgh). British attempts to dislodge the French were unsuccessful until the Forbes expedition of 1758 captured Fort Duquesne. As the French retreated northward, Fort Le Boeuf became more vulnerable to scouting attacks and small skirmishes. In August 1759 the French were forced to abandon Le Boeuf and, in their retreat to Detroit, burned the fort.

The English, realizing the necessity of guarding their newly gained frontier, sent Colonel Henry Bouquet north from Fort Pitt in October 1760 to build an English fort at Le Boeuf. This was occupied by the British until an Indian attack during Pontiac's Conspiracy forced the small garrison to Fort Pitt in June 1763. Once again, Fort Le Boeuf was burned.

The British did not rebuild the fort. It was not until thirty years later that it was rebuilt. With the end of the Revolutionary War, the Indian danger necessitated the reestablishment of a defense system, and in 1794 Governor Thomas Mifflin of Pennsylvania authorized the construction of two small blockhouses under the command of Major Ebenezer Denny. General Anthony Wayne further strengthened frontier defenses by ordering the erection of a blockhouse at Le Boeuf in 1796.

Today, the foundation of the American fort may be seen together with the Judson House, a Greek revival home built in 1820 on the site of the original French fort. The house contains a small museum and a model of the French fort. Across the street, in a small park, are the remains of the 1796 American fort.

Fort Le Boeuf can be reached on U.S. 19 north from Meadeville or west on State 97 from Union City. Open from May to September, Tuesday to Saturday 9:00 A.M. to 5:00 P.M., Sunday 1:00 P.M. to 5:00 P.M. Adults 50¢. Children free.

Fort Ligonier

LIGONIER, PENNSYLVANIA

IN THE MIDDLE of the eighteenth century the European conquerors of the New World began to vie for the ownership of the vast inland basin of America. In 1749 the French sent a party down the Allegheny and Ohio rivers to claim the land for the King of France. To enforce this claim the French began to build a chain of forts from Quebec to the Mississippi, including Fort Le Boeuf, near the present site of Erie. Here, Washington came in December 1753 to deliver the British challenge. Finding that conflict was inevitable, the British decided to establish a key fort at the "Forks of the Ohio," as advised by Washington. This fort was hardly begun before the garrison surrendered to a superior force of French. British

"Bastions" of Fort Ligonier

The "King's Colors" flying over Fort Ligonier

arms were unavailing. Washington's unsuccessful stand at Fort Necessity and Braddock's tragic defeat in 1755 left the French in complete possession for the next three years.

Finally, in 1758 the British organized their strength to drive the French from the New World by simultaneous attacks on Quebec, Crown Point, Niagara, and Duquesne. General Forbes, who was assigned the task of taking Fort Duquesne, decided to abandon the Braddock route and extend the path westward through the forests from the recently completed Fort Bedford.

The distance to the Forks of the Ohio was too great for an army to travel without rest and reprovisioning. At the site of an Indian village called Loyalhannon, which lay almost halfway from Bedford, it was decided to build a fortified camp to serve as the "staging area" for the final assault. Work was begun September 4, 1758, on this post, later to be named Fort Ligonier in honor of Sir John Ligonier, commander in chief of the British Army.

By mid-October, Colonel Washington and his Virginia Regiment had joined the rapidly growing force there, but not till after the fort had already repulsed severe attacks by the French and Indians on October 12. After resting and reprovisioning, Forbes's army struck out and occupied Fort Duquesne on November 25, 1758. The French mined and burned the fort before they retreated.

Fort Ligonier was never taken by the enemy and served in history as the "Key to the West." It served as a safe refuge during Pontiac's War when it was attacked by the Indians in June of 1763. It withstood the siege, a distinction shared on the frontier only by Detroit and Fort Pitt. On March 20, 1766, the fort was decommissioned by General Thomas Gage.

Although all vestiges of the fort had disappeared more than 175 years ago, its plan and location have been accurately fixed from early military drawings and records, some of which are on display in the fort museum. Archaeological excavation has recovered several hundred objects which form the nucleus of the collection on exhibit. The original Fort Ligonier was built from earth and timber. Its design was based on principles long established in continental Europe. The inner fort was square, with pointed projections at the corners known as "bastions." Within this square were storehouses, officers' barracks, powder magazine, and other structures. The portion of the fort presently reconstructed is the lower half of that inner fort and includes the officers' barracks and two bastions, connected by a wall of palisades. Flying over the fort is the authentic flag of 1758, known as the King's Colors. It bears the Cross of St. George (England) and the Cross of St. Andrew (Scotland). The Cross of St. Patrick was added later when American and British authorities granted permission to display this flag.

Fort Ligonier is located on Lincoln Highway, U. S. 30, between Bedford and Pittsburgh. Route 711, south to Johnstown, also borders the site. Open April to November, daily from 9:00 A.M. to dusk. Adults $1.75. Children $1.00.

Fort Mott

SALEM, NEW JERSEY

IN 1837 the federal government was concerned for the safety of Philadelphia's port and decided to add to the ten-gun battery on the Delaware shore (Fort Dupont) and the gun battery on Pea Patch Island in the middle of the Delaware River (Fort Delaware). They purchased land for a battery on the New Jersey shore (Finns Point) as an auxiliary to the other two forts, but it wasn't until 1872 that the U. S. Army Corps of Engineers began building earthworks. During the Civil War a portion of the land on the fort site was used as a burial ground for Confederate prisoners of war who died in the prison camp at Fort Delaware on Pea Patch Island. Fort Delaware was known as "Andersonville of the North." For the record, there are monuments erected on Fort Delaware to the 2,436 Confederate and 300 Union soldiers buried there.

In May of 1878 active operations for establishing a permanent battery began with the erection and mounting of two eight-inch rifles.

Concrete emplacement and parapet, Fort Mott (Photo: New Jersey Dept. of Conservation & Economic Development)

No troops were stationed at Fort Mott at this time, and the work progressed intermittently until six emplacements were erected. The two rifles were the only armaments installed.

In 1896, on the eve of the Spanish-American War, work began on two batteries with a continuous parapet—Battery Harker and Battery Arnold. In 1897 a War Department Order by General Roder designated the new fort, Fort Mott, in honor of Major General Gersham Mott, a native of New Jersey who had a notable record in the Civil War and who died November 29, 1884. The battery armaments consisted of three twelve-inch rapid-fire guns and five Gatling guns. It was first garrisoned by Battery I, 4th Artillery, from Washington Barracks.

During the Spanish-American War the fort was garrisoned by the 14th, Pennsylvania Volunteer, and Battery L, 4th Artillery. In June of 1899 Battery H arrived from Fort Monroe in Virginia. The fort was regularly garrisoned thereafter until 1922 when a caretaking detachment remained in charge until October of 1943. In 1951 the state of New Jersey dedicated the site as a state park and it now occupies an area of 104 acres.

Today the site consists of gun mounts and underground fortifications. The promenade deck atop the revetments offers a vantage point to observe traffic on the Delaware River. A range-finder tower constructed in 1901 which directed the fire of the 12-inch guns as well as a tower for the ten-inch guns can be seen. The moat and barricades of earth, behind the batteries, protected the batteries from the rear. The brick building, near the moat, was the Peace Storage Magazine, built in 1904. This provided dry storage away from moisture of underground magazines. The track in the tunnel indicates the route to the guns. The one large white house is the only remaining residence once used for commissioned officers. Across the parade ground are two houses which were homes for noncommissioned officers. The crumbling walk north of the parade ground marks the location of barracks for enlisted men and the post hospital.

Fort Mott is at Finn's Point on the east bank of the Delaware River midway between Salem and Pennsville, Salem County, in New Jersey. Take the New Jersey Turnpike to Exit 1, turn south on Highway 49. It leads to Salem and the Fort Mott State Park. The park is open during daylight hours throughout the year. No admission fee.

Old Fort Mifflin

PHILADELPHIA, PENNSYLVANIA

"THAT CURSED LITTLE MUD FORT," snarled Lord Cornwallis, for the fort that was begun by the British in 1773 claims to be the shield that saved the American Revolution in October and November 1777 by delaying British troops long enough to allow George Washington and his defeated army to find a winter refuge at Valley Forge. Historians, however, tend to minimize this claim by stating that Sir William Howe, the British commander, found many reasons for not following Washington's troops to Valley Forge and that the delay caused by Old Fort Mifflin's resistance was probably a contributory one among many.

The battle of Old Fort Mifflin was one of the lesser-known but more vital battles of the Revolutionary War. But it was there that the

American troops who had been beaten at Germantown, Paoli, and Brandywine stood off British sea and ground forces with a handful of men. The fort's garrison blocked the Royal Navy in its efforts to get up the Delaware with supplies for British troops in Philadelphia. Opposite Old Fort Mifflin on the New Jersey side of the Delaware was Fort Mercer. Together the forts sealed off the large British fleet commanded by Howe's brother, Sir Richard, and left the twenty thousand British and Hessian troops in Philadelphia short of supplies.

The bombardment of the fort was, according to some local historians, "the mightiest bombardment of the 18th century." The British attack brought one thousand shots every twenty minutes. So much shot was flung at so few that the American artillery officers in the fort paid the men in rum for recovering British cannonballs. Many of the British shells burrowed into the deep mud inside the fort, snuffing out the fuses. The interior was a square of ooze crisscrossed by wooden walkways. Many of the British shells still may be under the fort. The river was blocked with submerged anchored logs pointed with iron spikes to snag British ships. These were designed by Benjamin Franklin. The channel on the other side of the island was closed by a chain. A forgetful American soldier or a British sympathizer left the chain off one night and British ships sailed up to the walls of the fort, firing at close range from cannons and muskets in the sail rigging, finally forcing American evacuation. An estimated 250 dead were left behind.

The fort was named for General Thomas Mifflin, who took over the British stronghold before the Revolution and supervised its strengthening with additional log walls in 1776 for the defense of Philadelphia by the Continental Army. Much of Old Fort Mifflin was erected in 1798 during President John Adams' administration. The original plans were drawn up by a French military engineer, Colonel Louis Toussard. In addition to the main building, other structures, including barracks and headquarters, were put up in the early 1800s and the 1860s. There are powder magazines and dungeonlike casemates in which Confederate prisoners were kept. In some of the casemates are furnaces where shot was heated, and in other dark passages are thick wooden tier bunks on which prisoners slept. Union deserters were executed on the parade grounds.

Visitors to the old fort, which is close to Philadelphia's International Airport, will see stout old walls circled by a moat. One enters through a large, heavily studded wooden door. An arch above has a date stone that reads, "A.D. MDCCXCVIII, John Adams, P. U.S." The original British stone wall still runs unevenly around the fort, and the lowest part is where the shelling did the most damage. The brick ramparts built on the British stone were laid during the 1798 reconstruction.

Present restoration includes the commissary, which has been turned into a visitor's center, and the blacksmith shop. Work has begun on the living quarters and the arsenal.

Old Fort Mifflin can be reached from central Philadelphia by taking the Schuylkill Expressway south and following airport signs to the Penrose Avenue Bridge. Open from 9:00 A.M. to 5:00 P.M. No admission fee listed.

Old Fort Niagara

YOUNGSTOWN, NEW YORK

DURING THE REIGN OF LOUIS XIV in the seventeenth century, it was the ambition of the noted explorer Robert Cavelier de La Salle to establish a vast empire for France in America. As a beginning of this ambitious undertaking, La Salle, carrying letters patent from the King, came to Quebec and organized an expedition of exploration and discovery.

Journeying southwestward, he crossed Lake Ontario, and in January of 1679, commenced construction of a crude fort at the mouth of the Niagara River, one of a chain of sixty forts which France eventually built to defend her claims. This was the beginning of Fort Niagara, "a cradle of civilization on the Niagara Frontier," and key to the great unexplored West.

La Salle's fort was destroyed by fire. In 1687 a second fort was built at the site by Denonville, royal governor of Canada. This was later abandoned. The present "Old Fort Niagara" was begun in 1726 with the construction by Gaspard Chaussegros de Lery, King Louis XV's chief engineer in Canada, of a fortified stone trading post, known today as the French "Castle." The Indians were deceived into believing it was to be merely a "stone house for trading," for they greatly outnumbered the French and their permission for its erection was felt necessary. As constructed it turned out to be one of the most strongly fortified buildings in America. The innocent-looking dormers on the top deck offered excellent gun platforms for six-pounder cannon, which, together with heavy rampart muskets, could pour a devastating fire on any assaulters.

During the ensuing years the rivalry between France and England for domination of the continent of North America and control of the fur trade grew in intensity. The fort at Niagara, enlarged and greatly strengthened by the skilled engineer Captain François Pouchot, the last French commander of Fort Niagara, was of strategic importance to the French, as it held open their lifeline of supplies, communication, and trade between Quebec and their fortified trading posts in Louisiana.

During the French and Indian War, Fort Niagara was placed under siege by General Sir William Johnson in 1759, and after a prolonged struggle, Fort Niagara was surrendered to the British. This severed lifeline was a principal factor in the collapse of French resistance in the west. Quebec fell a few weeks later, and the French and Indian War was virtually over.

Fort Niagara in British hands was further strengthened to resist attack. They built two massive stone blockhouses to afford a flanking protection of the French "Castle" and to afford a plunging fire into the outer approaches. The blockhouses and the "Castle" form a triangle.

During the American Revolution, Fort Niagara was a base of guerrilla warfare against the unprotected settlers of central New York and Pennsylvania. Thirteen years after the Revolution, in 1796, Britain relinquished Niagara and evacuated it under the terms of the Jay treaty. The fifteen stars and fifteen stripes were raised on its historic parade ground.

In the War of 1812 the old fort was again heavily engaged by the British. It was retaken by the British on December 19, 1813, but at the conclusion of the war it was returned to American hands and has remained so ever since. Today, Fort Niagara is an actively maintained historic shrine operated by the Old Fort Niagara Association. In addition to the two block-

Saluting the colors of the "Three Nations," Old Fort Niagra (Photo: Grove McClellan)

Drawbridge gate, Old Fort Niagra (Photo: Grove McClellan)

houses and French "Castle" the visitor can see the Bake House with the two original ovens intact, the hot-shot furnace which heated to red-hot heat the shot fired from the cannon on the gun deck of the "Castle," the Powder Magazine, the drawbridge, and the Gate of the Five Nations which formed the main entrance to the fort, cannon, artifacts of the daily life at the fort, historic flags, etc.

Old Fort Niagara is fourteen miles north of Niagara Falls. Leaving the Falls, take the Robert Moses Parkway north through Lewiston and Youngstown to the fort. Open daily from 9:00 A.M. to dusk. Admission $1.00. Children under twelve free.

Fort Stanwix

ROME, NEW YORK

THE STAND by an American garrison at Fort Stanwix during August 1777 was chiefly responsible for the repulse of the western wing of the British invasion of the northern colonies from Canada, and checked the possibility of a loyalist uprising in the Mohawk Valley. The retreat to Canada of the western column after its failure to take Fort Stanwix was a blow to the British strategy of concentration at Albany, contributing thereby to the defeat of Burgoyne at Saratoga a few months later. In addition to its role in the War for Independence, Fort Stanwix was the scene of the treaty of that name, signed on November 5, 1768. By the Treaty of Fort Stanwix the Iroquois ceded a vast territory, south and east of the Ohio River, as far west as the mouth of the Tennessee. The treaty thus cleared the way for a new and significant surge of westward settlement.

Fort Stanwix was situated at the Oneida Carrying Place, a key spot on the route between the Great Lakes and the Mohawk River. It was built originally during the French and Indian War but played no significant part in this conflict. It was reestablished in June 1776 (sometimes called Fort Schuyler by the patriots) and garrisoned with perhaps as many as eight hundred men in time to block the British invasion objectives in the Mohawk Valley in the summer of 1777. General John Burgoyne advanced south from Canada along the Champlain route at this time, expecting to meet the main British Army under General Howe which he believed would move up to the Hudson. Colonel Barry St. Leger with more than one thousand regulars, Tories, and Indians was to move down the Mohawk Valley to Albany and join the larger British forces there after rallying Tories and Indians on his route.

St. Leger invested Fort Stanwix on August 3 but was rebuffed when he demanded its surrender. The action was limited to sniping until August 6, when the bloody battle was fought at Oriskany, some six miles to the east between St. Leger and an American militia force under General Nicholas Herkimer. The patriots were badly mauled and did not succeed in raising the siege at Stanwix, but during the action a detachment from the fort raided the British position, destroying provisions and camp equipment. This encouraged the besieged, who held firm while St. Leger began formal siege operations. He had advanced his works to within 150 yards of the fort when word came of the approach of an American relief force under General Benedict Arnold. Having lost the confidence and support of his Indian allies, St. Leger

Fort Stanwix, named for the British general responsible for building it in 1758, was an American garrison during the Revolutionary War. (Photo: Richard Frear)

Costumed militiamen guard the moat-spanning drawbridge outside the log walls of reconstructed Fort Stanwix (Photo: Richard Frear)

18

was obliged to abandon the siege near the end of August, retiring in considerable disorder to Canada. Fort Stanwix still stood, and the American Army on the Hudson could give its full attention to Burgoyne, who surrendered at Saratoga on October 17, 1777.

The site of Fort Stanwix occupies approximately a city block in the heart of Rome, New York, and no physical evidence of the post is visible. The site is built over with roads, houses, and commercial developments. The remains of the fort were cleared away prior to the middle of the nineteenth century. The fort has been completely reconstructed by the National Park Service.

Fort Stanwix site is in downtown Rome bounded approximately by Dominick, Spring, Liberty, and North James streets. It is open from June to Labor Day, Monday to Friday, 9:00 A.M. to 4:00 P.M., Saturday from 9:00 A.M. to 7:00 P.M., Sundays from 12:00 noon to 6:00 P.M. No admission fee.

Fort Ticonderoga

TICONDEROGA, NEW YORK

STRATEGICALLY LOCATED at the junction of Lake Champlain and Lake George, Fort Ticonderoga was the key to both Canada and the Hudson Valley in the eighteenth century. It saw more of the English-French struggle for North America than any other post, and its story is one of the most dramatic and colorful in American military annals.

The first military post on the site was Fort Vaudreil, later Fort Carillon, built by the French in 1755-57. On July 8, 1758, an army of fifteen thousand British regulars and colonial troops attacked the fort and was repulsed with heavy losses by the French under Montcalm. On July 27, 1759, however, General Jeffrey Amherst captured the fort and renamed it Ticonderoga. This loss by the French, coupled with British pressure elsewhere on the frontier between New France and the American colonies, was a severe blow to French plans. The capture of Ticonderoga gave the British undisputed possession of the strategically important Hudson River Valley. The French blew up part of the fort before they withdrew, and Amherst had repairs made in accordance with the original design. In the years between the defeat of France in North America and the outbreak of the Revolution, a small garrison manned the work. On May 10, 1775, Ethan Allen with eighty-three "Green Mountain Boys" surprised and defeated the few British defenders, and the post became a base for the projected advance on Canada. The following winter Colonel Henry Knox hauled the fort's cannon overland to serve in the siege of Boston. Ticonderoga changed hands again when it fell to Burgoyne's British Army in the summer of 1777, but upon Burgoyne's defeat at Saratoga it again passed into American possession. Although reoccupied from time to time by scouting parties and raiding detachments, the post was never again garrisoned by a military force.

In 1816 William F. Pell, a merchant of New York, leased the grounds and four years later bought them. In 1908 the late Stephen Pell began restoration. By the following year the west barracks had been opened to the public, and the work has gone forward since that time. The task of reconstruction was a major undertaking. Over the years the stones had been carted away by settlers for use as building

Air view, Fort Ticonderoga, looking south up Lake Champlain, Mount Independence in background (Photo courtesy of Fort Ticonderoga Museum)

South wall and south barracks, Fort Ticonderoga (Photo courtesy of Fort Ticonderoga Museum)

materials. The upper part of the walls and most of the stone barracks disappeared, and the earth behind the walls washed over the remnants of the original walls. These remains were uncovered in the restoration that began in 1908. The present work was erected on the original foundations and utilized parts of the walls that had survived.

The fort is four-sided with bastions extending from its four corners. Outlooks or demi-lunes on the north and west, and an outer wall on the south, cover the approaches. Facing the central parade ground are the reconstructed west and south barracks, the east barracks, and the long rampart joining the northwest and northeast bastions. The west barracks houses the administrative office, a library, and, in the basement, the armory, featuring the most important part of the Fort Ticonderoga gun collection. In the south barracks are displayed many artifacts excavated in the course of the restoration; furnished quarters of the officer of the day; exhibits of furniture, household goods, and other items used by early settlers of the region; Indian relics; and a model of the fort as it existed in 1758. Below the walls are the remains of a French village that probably served the fort.

Fort Ticonderoga is in Essex County on New York 8 and 9N in Ticonderoga, New York. It is owned and administered by the Fort Ticonderoga Association. It is open from mid-October to mid-May. Adults $2.00. Children $1.00.

Fort William Henry
LAKE GEORGE, NEW YORK

THE YEAR 1755 found France and England fighting for supremacy in North America. General William Johnson was given command of a small army of American provincial troops and Mohawk Indians. They assembled at Albany and began a northward march along the Hudson in a campaign aimed at driving the French from their positions at Crown Point on Lake Champlain. On August 28, 1755, Johnson, with a vanguard of fifteen hundred men, reached the head of Lake St. Sacrement, which he rechristened Lake George in honor of George III.

In the meantime a French expedition under Baron de Dieskau had set out from Canada to crush Johnson's army. Early in September he advanced to Ticonderoga. He left part of his force at Ticonderoga and embarked in canoes on Lake Champlain with 216 French regulars, 684 Canadians, and 600 Indians. Landing at the head of South Bay, he marched overland. The first step of his campaign was to surprise the garrison at Fort Lyman. To his dismay he discovered that the forces he faced at Lyman were formidable and armed with cannon. The sight of the artillery threw the Indians into a panic and they refused to attack the fort but after much persuasion agreed to assault Johnson's unprotected position on Lake George. Dieskau changed his strategy and moved on Johnson's army.

Johnson's Indian scouts reported movements of Dieskau's forces, and on the morning of September 8, 1755, he dispatched one thousand men under Colonel Ephraim Williams to intercept the French. But unknown to both Johnson and Williams, one of their men was captured by Dieskau as he neared Johnson's camp. The prisoner told Dieskau of the advancing troops under Williams, and Dieskau prepared an ambush. He concealed his Indians and Canadians on either side of the road and placed his regulars up front and awaited his unsuspecting enemy. As Williams' men reached Rocky

Brook, some three miles from Johnson's camp, Dieskau's troops poured volley after volley into them. The only event that prevented the trap from being a complete surprise was that Dieskau's Senecas recognized their Mohawk brethren in the advancing column and fired into the air to warn them. In the battle among the rocks and trees Williams was killed and the survivors retreated back to camp on the lake.

Johnson, at Lake George, hearing the firing from Rocky Brook, threw up a barricade around his position. The survivors of the ambush reached the camp followed by the pursuing French. Dieskau assaulted Johnson's works, but the Iroquois, allied to the French, again retreated from the feared cannon in the English camp. The French regulars were unable to penetrate the fire of the English. After several hours, the colonials leaped over their breastworks and counterattacked, routing the French, who suffered heavy losses, and taking the wounded Baron de Dieskau prisoner.

The sounds of battle also attracted Colonel Blanchard in command of Fort Lyman and he dispatched 250 men in support of Johnson. This unit came upon several hundred Canadians and Indians having their lunch from their packs at Rocky Brook, the scene of the ambush. The colonials attacked and killed or scattered their adversaries. The bodies of the slain were thrown into the water, which was known thereafter as "Bloody Pond." And so the battle of Lake George ended. Both sides lost over two hundred men. This was the first battle fought wholly by American provincials against trained European troops.

Johnson remained at Lake George through the autumn where he built Fort William Henry, named after one of the grandsons of King George. The soldiers built the fort. They set to work laying logs on the prominence. The surrounding country was cleared of trees and the logs brought into the entrenched camp where they were squared down and moved into place for the new fort. Thus they cleared a range for artillery and provided material for the new fort. In two weeks' time the outer walls of a four-bastioned fort were completed. The Militiamen began the construction of underground magazines, storehouses, and casemate rooms. On September 10, 1755, Fort William Henry was completed.

In March 1757 Rigaud Vaudreuil, with 1,600 French regulars and Canadian militia and Indians from Fort Carillon, approached Fort William Henry over the frozen surface of the lake to take the fort in a surprise attack. The garrison of 346 British troops and rangers were alerted and met them with stiff resistance. The French laid siege for three days, burning some storehouses and some boats frozen in the ice, and then retired. There were subsequent skirmishes between the French and Indians and scouting parties operating from Fort William Henry. The end of the fort came on August 3, 1757, when an army of 6,000 French regulars and Canadians and 1,700 Indians under the command of the Marquis de Montcalm appeared before Fort William Henry, having traveled up the lake from Ticonderoga. He promptly demanded the surrender of the fort, but the English commander, Colonel George Monro, expecting help from Fort Edward, refused. Monro's command consisted of 500 men in the fort and 1,700 in an entrenched camp southeast of the fort. Montcalm began a bombardment of the fort and each day moved closer with his guns. Soon half of Monro's guns were silenced. He held out against the overwhelming French forces until August 9, when, with more than 300 of his men killed or wounded and others sick with smallpox, he agreed to surrender. A condition of surrender was that he and his men, together with their women and children, were to be safely escorted to Fort Edward. On the morning of August 10, as the

Replica of third fort built on this site; original built in 1692 (Photo: Norton, Maine Department of Economic Development)

garrison marched out of the camp on the road to Edward, Indians fell upon the column with their tomahawks, and a bloody massacre followed. Some 50 men, women, and children were killed and 200 carried off to Montreal as prisoners of the Indians. Montcalm and his officers rushed to the scene to bring a halt to the slaughter. Montcalm burned Fort William Henry on August 16.

In 1758 General Abercrombie moved into the head of Lake George with a very large army. On the site of the burned-out fort they threw up an earthwork and armed it with several pieces of small-caliber cannon. During the Revolution, archaeological excavations indicated that there was a considerable encampment of British and Tory troops inside the fort ruins.

In the fall of 1952 it was decided to preserve the earthen mounds that constituted the physical remains of Fort William Henry and to erect a replica of the original fort. Canadian loggers were sent into the mountains for timber. Efforts were made to exactly locate the Military Cemetery. Plans of the original fort were obtained from the British Archives, the Canadian Archives, the Library of Congress, the New York State Museum, and the Clements Library. Artifacts were recovered in the diggings, and little by little the reconstruction of Fort William Henry was completed.

Fort William Henry and a reconstruction of an Iroquois Village can be visited on Canada Street in Lake George, New York. It is open from May to July daily from 10:00 A.M. to 5:00 P.M. From July to August daily from 10:00 A.M. to 10:00 P.M. From September to October daily from 10:00 A.M. to 5:00 P.M. Adults $2.25. Children $1.00.

New England

Fort Halifax

WINSLOW, MAINE

FORT HALIFAX is believed to be the oldest wooden blockhouse in the United States. It was constructed at the confluence of the Kennebec and Sebasticook rivers in 1754, "the only known communication which the Penobscots have with the River Kennebec and the Norridgewack Indians is through the Sebasticook . . . a fort here cuts off the Penobscots not only from the Norridgewacks but also from Quebec," runs a contemporary account of the fort.

The original grant to the land around Fort Halifax goes back to King James I of England, who, in 1606, issued patents to the London and Plymouth companies. Where he obtained the rights to this land from the Indians is not known.

Fort Halifax Memorial. Located in Winslow, it is the oldest original wooden block house in the U.S.

26

The Kennebec River was visited by Edward Winslow as early as 1625 for the purpose of exploring the potential of trade with the Indians. In 1628 a trading post was built twenty miles below at Cushnoc (Augusta). The beaver skins from the Abnaki Indians of the Kennebec Valley contributed greatly toward paying the Pilgrims' heavy debts in England.

In 1629 Governor Bradford and others were given a large tract, later deeded to the colony and sold in 1661. For eighty-eight years things were quiet along the Kennebec. In 1749 the Proprietors of the Kennebec Purchase bought the land from the descendants of the 1661 purchasers. The proprietors were land developers, and four of them gave their names to nearby towns—Gardiner, Bowdoin, Vassalboro, and Hallowell. Fort Halifax, named for the Earl of Halifax, Secretary of State for the Kingdom of Great Britain, was built to encourage settlers and give some assurance of safety.

On April 16, 1754, Governor Shirley wrote to the Plymouth proprietors, "I shall order a house of hewn timber not less than 10" thick, 100 feet long and 32 feet wide and 16 feet high, for the reception of the province's stores with convenience for lodging the soldiers . . . and build a blockhouse 24 feet square agreeable to a plan exhibited by you to me . . . to cause the same to be finished with the utmost expedition."

The first officer in charge was General Winslow, for whom the town was named, who came with eight hundred men in the spring of 1754. Three hundred remained the summer to complete the complex of two blockhouses, a sentry house, a barracks, and a large building. The last was used for officers' quarters, storehouse, church services, and town meetings. The first winter was one of hardship and there were skirmishes with the Indians.

In 1775 Colonel Benedict Arnold with eleven hundred men stopped here on his way to Quebec. At that time the large house within the fort was used as a tavern. Afterward it was used as a dwelling house, meeting house, town hall, and finally a home for poor families until torn down in 1797.

Through the years the buildings have been torn down with the exception of the blockhouse. In 1913 the Winslow Chapter of the DAR was organized. In 1924 they acquired title to the fort and maintained it until 1965, when they deeded it to the state of Maine.

Fort Halifax is located on U.S. 201 and State Highway 100, one mile south of Winslow-Waterville Bridge. It is open from May 30 to Labor Day. Admission fee not specified.

Fort Knox

PROSPECT, MAINE

FORT KNOX was named for Major General Henry Knox of Thomaston, Maine. He served as an adviser during the Revolutionary War to General George Washington and later as his Secretary of War. The site of the fort was chosen as a fortification at the time of the controversy over the northeastern boundary. During this dispute there was great fear along the Penobscot that the British might invade Maine by way of the river.

Construction of the fort began in 1844 and stopped in 1864 before being completed. It was constructed of granite from Mount Waldo, a few miles up the Penobscot River. The mason-

The cannon balls were heated in this furnace. Red hot, they could skip on the water and yet set a ship afire.

Heating a 24 pound ball took about 25 minutes.

Fort Knox Memorial, named for General Henry Knox

ry in the fort is the work of master craftsmen, withstanding the elements for a century and remaining in excellent condition. The underground stairways, the curved brick arches, and the circular stairs inspire a deep respect for the workmanship of these vanished craftsmen. This is one of the largest forts of this type of construction in the country.

Fort Knox was first occupied by a detachment of Maine volunteers from July 1863 to March 1866. The monthly post returns show that during the Civil War the garrison consisted of about fifty men. During the Spanish-American War (June and July of 1898) Fort

Knox was occupied by Connecticut volunteers.

The State of Maine purchased Fort Knox from the U. S. Government on October 12, 1923, for $2,121. It was assigned to the State Park Commission on July 1, 1943.

The most interesting features that a visitor can inspect are the solid old dark stairways and passageways and the huge brick arches. Fort Knox can be reached on U.S. Route L from Bucksport or Stockton Springs, or Route 174 from Prospect. Open from May to November 1, daily from 10:00 A.M. to 6:00 P.M. Parking fee $1.50.

Old Fort #4

CHARLESTOWN, NEW HAMPSHIRE

ORIGINAL FORT #4 dates back two and a quarter centuries. It was the northernmost outpost in the wilderness of that day, and the settlers of the area looked to its protective logs to help them "keep their scalps on" against Indian marauders.

Two hundred and twenty years ago, thirty farmer-soldier defenders under the command of Captain Stevens fought off a band of seven hundred Indians who tried to destroy the bastion by fire and by shot. For his valor in defending Old #4, Captain Stevens was presented a ceremonial sword by the colonial government, and the attackers never again mustered the strength to mount an assault of anything approaching equal fury.

The original bastion has long been gone. It was located on what is now Charlestown's wide Main Street, site of several lovely colonial homes. Archaelogical probings have pinpointed

the exact location of the original fortification.

Reconstruction of the fort and erecting almost a dozen cabin-type structures is completed. Some half-million board feet of lumber has gone into its construction. The stockade consists of seven hundred logs, separated by a very few inches, as was the original. This was so defenders could poke the barrels of their muskets between the logs and get a maximum field of fire. All of the logs are native New Hampshire pine.

Old Fort #4 is located in Charlestown on the westerly side of Route 12 midway between the business and residential sections of Charlestown. Open from June 13 to Labor Day, weekdays from 11:00 A.M. to 5:00 P.M., Saturday and Sunday from 11:00 A.M. to 5:00 P.M. Adults $2.00. Children from seven to twelve 75¢; under seven free.

Old Fort #4. Charlestown, New Hampshire (Photo: Office of Vacation Travel, Concord, New Hampshire)

Plimoth Plantation

PLYMOUTH, MASSACHUSETTS

SOMETIMES between 1622 and 1623 the Pilgrims, after an incredible journey from England, in a tiny cockleshell of a ship named the *Mayflower*, which lasted sixty-six days across the stormy Atlantic, landed on the New England coast and established a settlement which they called Plimoth Plantation. A strong fort was erected to protect the twenty or so houses of the community. This was one of the earliest of American forts.

An account written in 1623 describes the town and its fortification as follows: "It [Plymouth] is well situated upon a high hill close unto the seaside, and very commodious for shipping to come unto them. In this plantation is about twenty houses, four or five of which are very fair and pleasant, and the rest (as time will serve) shall be made better. And this town is in such manner that it makes a great street between the houses, and at the upper end of the town there is a strong fort, both by nature and art, with six pieces of reasonable good artillery mounted thereon; in which fort is continual watch, so that no Indian can come near thereabouts but he is presently seen. The town is paled around about with pale of eight feet long, or thereabouts, and in the pale are three great gates."

Isaack de Rasieres, Secretary to Peter Minuit, governor of New Netherlands, wrote a letter in 1627 to an Amsterdam merchant describing the town and fort: "New Plymouth lies on the slope of a hill stretching east towards the sea-coast, with a broad street about a cannon shot of 800 feet long, leading down the hill; with a [street] crossing in the middle, northwards to the rivulet and southwards to the land. The houses are constructed of clapboards, so that their houses and courtyards are arranged in very good order, with a stockade against sudden attack; and at the ends of the streets, stands the Governor's house, before which is a square stockade upon which four patereros are mounted, so as to enfilade the street. Upon the hill they have a large square house, with a flat roof, built of thick sawn planks stayed with oak beams, upon the top of which they have six

cannons, which shoot iron balls of four and five pounds, and command the surrounding country. The lower part they use for their church, where they preach on Sundays and the usual holidays. They assemble by beat of drum, each with his musket or firelock, in front of the Captain's door; they have their cloaks on, and place themselves in order, three abreast, and are lead by a sergeant without beat of drum. Behind comes the Governor, in a long robe; beside him comes the preacher with his cloak on, and on the left hand, the captain with his side-arms and cloak on, and with a small cane in his hand; and so they march in good order, and each sets his arms down near him. Thus they are constantly on their guard night and day."

The Fort and the community have been re-created using old records and eyewitness accounts of visitors to the original Pilgrim colony, archaelogical research, and the history written by their leader, William Bradford. Every effort was made to create a functioning village set in the year of 1627. Guides and hostesses in Pilgrim dress carry on the tasks necessary for living in a seventeenth century farming community. A replica of the *Mayflower II* is also on view. There are replicas of the cannon in the reconstructed stockade. The entire re-creation of Plimoth Plantation is about two and a half miles below the original site.

Plimoth Plantation is about three miles west of Plymouth, Massachusetts. State Highway 3A out of Plymouth leads to the fort and village. The site is open from April to November, daily and on holidays from 10:00 A.M. to 5:00 P.M. During the summer months it is open from 10:00 A.M. to 6:00 P.M. Adults $1.50. Children 50¢.

Street of Restoration, Plimoth Plantation

Tilling field, Plimoth Plantation

Fort Western

AUGUSTA, MAINE

BUILT IN 1754, Fort Western is the only fort still standing in Maine that predates the Revolution. It was built as a stronghold by the Plymouth Proprietors at Cushnoc, the head of navigation on the Kennebec, to hold supplies for shipment by oxcart to Fort Halifax at Taconic Falls, now Winslow, built by the British Army in the same year. One original blockhouse of Fort Halifax still remains. Fort Western is unique in that it has served as a trading post, barracks for soldiers, and also living quarters for Captain James Howard, the first and only commander of the fort, and his family. It has much of its original material in the restoration.

Gersham Flagg of Boston, one of the Proprietors of the Kennebec Purchase, who was a housewright and glazier by trade, directed the building of the fort. Hand-hewn logs were cut and dovetailed at Topsham and floated up the river on the tide, a measure of precaution that was necessary because of fear of Indian attack. The fort was built on a stone foundation, contained twenty rooms, and had seven flights of stairs. The main building now standing is one hundred feet long, thirty-two feet wide, and sixteen feet high. Its solid walls are twelve inches thick.

Long before the fort was completed, James Howard, then a lieutenant, was placed in command of twenty men stationed there. His family soon followed him to live at Fort Western. James Howard was the largest landowner at that time and the most influential man in the little frontier settlement. When the fort was first built it was fitted with officers' quarters at each end with a large space left for storage in the center of the building. This central space afterward became the "trucking house" and store, conducted by Captain Howard and his sons, who brought supplies for the settlers by boat from Boston.

Because of possible Indian attack, no attempt was made to settle the country about the fort until after the fall of Quebec in 1759, and no houses were built at Cushnoc outside the palisade until then. The fort was never attacked but retained its garrison because of unsettled Indian affairs, and as late as January 1764, Governor Bernard recommended to the General Court that the fort be maintained. In 1762 the garrison was reduced to one lieutenant, one armorer, two sergeants, and thirteen privates.

In the early days of the settlement, all public meetings were held at the fort. The first public religious service was held there and the first marriage was solemnized there in 1763, when Margaret Howard was married in a ceremony officiated by her father, who, as justice of the peace, was the only person in the settlement qualified to perform the ceremony.

In 1763, after obtaining large grants of land at Cushnoc, Captain Howard built a "great House" about a mile above the fort and lived there until his death in 1787. In December 1769 the Proprietors of the Kennebec Purchase sold Captain Howard Fort Western and about nine hundred acres of surrounding land. The fort then became a trading post.

The role of Fort Western in the American Revolution was an interesting, if minor, one. General Benedict Arnold, controversial even in the early days of the war, launched his ill-timed march to Quebec from Fort Western in the fall of 1775. Together with eleven hundred men, Aaron Burr among then, Arnold assembled his supplies at the fort and then moved northward through the Maine woods to attack the British

Building with blockhouses, Fort Western (Photo: Robert D. Hotelling)

at Quebec. Disease, hunger, snowstorms, all took their toll of the badly decimated army that arrived before the walls of Quebec. Their attack failed and the survivors were forced to return to the United States. In 1779 the survivors of a Massachusetts expedition, sent out by Massachusetts to dislodge a British force at Castine, stopped at Fort Western on their way home. The expedition was ill managed. Paul Revere was artillery commander of this party.

Descendants of the Howard family remained in residence at the fort until it passed out of the family name and was divided up and occupied as a tenement house. Surrounded by decrepit buildings, in time it decayed. It was a refuge for the illegal sale of liquor, a fire risk, and a menace to the city. In 1919, with the aid of experts in historic preservation, the building was restored. In 1921 it was presented to the city of Augusta. The blockhouses and stockade were reproduced at this time, but the main building is original and restored.

The central part of the fort is composed of a trading room, kitchen, mess room, and barracks, while in the north and south sections of the building are the sitting rooms, bedrooms, and kitchen of the Howard families. The historical collections include naval and military articles, Indian and Western collections, as well as a room devoted to the craft of spinning and weaving.

Fort Western is located on Cony Street, fronting the Kennebec River in Augusta, Maine. It is open from May 15 through Labor Day, weekdays from 10:00 A.M. to 5:00 P.M., Sunday from 1:00 to 5:00 P.M. No admission fee.

34

Midwest & West

Fort A. Lincoln

MANDAN, NORTH DAKOTA

FORT LINCOLN was the headquarters of Brevet Major General George A. Custer and his famed 7th Cavalry. It was from here that he and his regiment set out on their Black Hills Expedition in 1874 which was a prelude to the famous "Black Hills Gold Rush." It was also from Fort Lincoln that the ill-fated Little Bighorn Expedition departed on May 7, 1876, during which Colonel Custer (he was promoted to Major General posthumously) and his entire command of 265 officers and men, many of whom were from Fort Lincoln, were annihilated on June 25. Several weeks later, the steamer *Far West* brought the tragic news of the disaster to the wives and families at the post.

7th Cavalry muster at Fort Lincoln State Park (Photo: North Dakota Travel Photo)

36

Fort Lincoln was built about 1873 and 1874. It was built with quarters for six companies. The barracks for the soldiers were on the side of the parade ground nearest the river, while seven detached houses for officers faced the river opposite. On the left side of the parade ground was the long granary and the military prison. Opposite the garrison proper were the stables for six hundred horses. Some distance on were the log huts of the Indian scouts and their families.

While the fort was under construction there was constant trouble from the Indians. War parties of Sioux occasionally raided the mules, horses, and cattle. As late as July 1876, the Indians drove off cattle under the eyes of the fort.

The fort was far removed from white settlements. The officers and men adjusted to frontier living. There were parties, dances, masquerades, and the men hunted deer, wolves, and coyotes with packs of hounds. The Sutler's Store operated a billiard parlor for the officers, and sleighing was a favorite pastime during the winter months.

The enlisted man found recreation in the grog shops at "The Point," which was on the east bank of the Missouri. Each of the companies gave a ball at which all attended. The enlisted men organized a Fort Lincoln Dramatic Association, which held its entertainments in a theater built in the cavalry barracks. Sometimes the Indians danced for the officers and enlisted men.

For several years following the Battle of Little Bighorn, the post was very active. In the fall of 1876 General Alfred Terry, after returning from the Little Big Horn expedition, organized a force of twelve hundred men at Fort Lincoln and proceeded to disarm the Indians in the various agencies on the river and confiscated their ponies. The following year the 7th cavalry left Fort Lincoln for the Yellowstone country, where it participated in the Nez Perce War. After work was renewed on the Northern Pacific Railroad in 1879, the fort gave protection to the various surveying and construction parties.

During the middle 1880s the fort declined rapidly. With the completion of the railroad in 1883, there was little need for its services. In 1889 the ordinance depot, which had been established at the fort in 1878, was moved to Fort Snelling, Minnesota. The country was rapidly filling with settlers, and in 1889 North Dakota was admitted to statehood. No longer needed, the post was ordered to disband. On July 22, 1891, Fort Lincoln was abandoned. Following its abandonment several attempts were made by the state to use the buildings. None of these were permanent. On December 1, 1894, about one hundred men with tools and teams dismantled the buildings illegally. When the marauders finished, only three buildings remained. Today, Fort A. Lincoln is a historic site in Fort Lincoln State Park. The blockhouses and a unique restored Indian earth-lodge village and a museum can be seen.

Fort A. Lincoln is several miles south of Mandan on Highway 6. There are facilities for trailer camping, boating, and fishing. The fort is open May and September to October, daily from 9:00 A.M. to 5:00 P.M.; June, daily from 9:00 A.M. to 9:00 P.M. Adults 25¢. Children under six free.

Fort Abercrombie

ABERCROMBIE, NORTH DAKOTA

Built on August 28, 1858, to protect the northwestern frontier and to become a link in a chain of military posts which extended along the route from St. Paul to the Montana gold fields, Fort Abercrombie was located on the west bank of the Red River of the North and established by Lieutenant Colonel John J. Abercrombie.

The most dramatic role that Fort Abercrombie played, however, was during the Sioux Uprising in Minnesota during 1862 when it was besieged for almost six weeks by the Sioux. Regular troops were withdrawn to the Southwest because of the Civil War, and the post was garrisoned during the siege by seventy-eight men and Company D, 5th Minnesota volunteer Infantry, under the command of Captain John Vander Horck. At that time the fort had no stockade but consisted of a drab group of scattered buildings—two wooden quarter buildings, one of which served for officers and the other for men, a small brick commissary stores building, a small guardhouse, a combined sutler's store and post office, and four stables.

Despite reenforcements, the Indians continued their siege. On the night of September 29, the Indians attacked a horse-watering detail, wounding a teamster. Several shells from the fort dispersed the Indians. This engagement ended the long siege of the fort, in which the casualties of the post were five dead and five wounded. Following the siege, the fort took immediate steps to repel any further attacks from the Indians. The timber and brush were cleared to prevent the Indians from creeping up close undiscovered and the construction of blockhouses and palisades was completed by February 1863, although no stockade was placed on the river or east side.

For several years Fort Abercrombie played an important part in the history of the northwestern frontier. In 1863 the Fisk Expedition, on their way to the gold fields in Montana, stopped at the post where the members repaired their wagons and shoed their horses and mules. In the same year the sick and wounded from the General H. Sibley Expedition were brought to the fort. The post at times furnished escorts for settlers and policed the region generally. Because of its location, supplies, mails, and troops bound for Fort Ransom, Wadsworth, Pembina, Totten, Cross, and sometimes to the Missouri River posts passed through Fort Abercrombie en route to and from St. Paul.

The military reservation surveyed in 1867 contained twenty-seven square miles on both sides of the Red River. Built to accommodate three companies of men, the average number of officers and men stationed there from 1867 to 1877 was one hundred. In 1875 stages ran three times a week from Breckenridge to Fort Abercrombie, Garry, Pembina, and Moorhead.

In common with other military posts white settlements pushed westward, and eventually Fort Abercrombie had served its usefulness. It accordingly was abandoned and the troops withdrawn October 23, 1877.

Today, the original log guardhouse remains on its site, and palisades and three blockhouses are the beginning of the reconstruction of the old fort. It is a State Historical Park, and trailer camping is welcomed. Fort Abercrombie is south of Fargo on Highway 81. It is open from May to October, Monday to Saturday from 9:00 A.M. to 9:00 P.M.; Sunday from 10:00 A.M. to 9:00 P.M. Adults 25¢. Children 10¢.

Fort Atkinson

WINNESHEIK COUNTY, IOWA

FORT ATKINSON was built so that a tribe of Indians might find a safer and better life. To understand the reason for building the fort, it is necessary to know something of the Indian treaties of the fifteen years preceding the fort, which was built in 1840.

In 1825 a great council was held between warring Indian tribes and the federal govern-

Foundation and reconstructed building, Fort Atkinson

ment in an effort to bring peace. A "neutral line" was established from the mouth of the upper Iowa River near New Albin. It extended to Hawarden on the Big Sioux River and thence down the Missouri River.

This neutral line did not stop all of the fighting. The Sioux to the north and the Sac and Fox to the south ignored it and fought intermittently. Other peace councils held for the next three years also failed to stop the bloodshed, causing national concern.

In the council of 1830 the "neutral ground" was established (a strip forty miles wide having as its center the previously established neutral line). This buffer strip also failed, for when Chief Black Hawk's band sought refuge within it, the Sioux fell upon them with horrendous results.

In the treaty of 1832 the Winnebago ceded all their land east of the Mississippi to the government and agreed to be moved to "neutral ground." The government was to provide them with agricultural implements, establish schools, and pay the tribe ten thousand dollars a year for twenty-seven years. However, only small bands of Winnebago actually moved in because the Sac, Fox, and Sioux tribes were hostile to them and the Sioux regarded this ground as their best hunting area.

No attempt was made to force the Winnebago to move until 1840, when Brigadier General Henry Atkinson, commandant at Jefferson Barracks, St. Louis, ordered troops to aid those at Fort Crawford, Prairie du Chien, if the tribe resisted. After many parleys, meetings, and councils, the move was effected as far west as the Turkey River. The Winnebago feared the other tribes and they also wanted to be near the trading center of Prairie du Chien.

General Atkinson had previously recommended the establishment of an army post on the Cedar River but changed his plan and on May 5, 1840, sent a party east to locate a good

spot for the post. On May 31, 1840, Captain Isaac Lynd with seventy-one men and officers of the 5th Infantry arrived at Fort Atkinson.

By the spring of 1841, it was clear that the foot soldiers could not cope with the restless Winnebago. Small bands were constantly returning to Wisconsin. Permission was then given to build stables and quarters for cavalry. On June 24, 1841, Company B, 1st Regiment of Dragoons, arrived at the post to do most of the patrolling necessary to keep the Indians within bounds.

Although Fort Atkinson was heavily armed, there is no record of its ever having been attacked. Minor revolts and skirmishes were ended quickly with the help of Morgan's Company of Iowa volunteers, who furnished their own horses and rode with the Dragoons for fourteen months. They were of much help when the Winnebago moved to Minnesota in 1848, by treaty, to a new reservation at Fort Snelling.

Fort Atkinson was no longer needed. The last troops left on February 24, 1849. Its supplies were sold and a caretaker was appointed to protect the buildings and grounds. The stockade was used for firewood by travelers. Settlers building homes found the fort a fruitful source of glass and hardware.

Meanwhile, Iowa had become a state, and the legislature was petitioning the federal government to give the fort and two sections of land to the state for the establishment of an agricultural college. The general assemblies of 1849-51 and 1853 urged their senators and representatives to use their influence to secure this land grant, but the War Department never even seriously considered it.

Buildings at the fort were sold in 1855 for about three thousand dollars and the land was given to the General Land Office for disposition and regular entry of the land by settlers for $1.25 per acre. All of the buildings were destroyed except the southwest blockhouse, the powder magazine in the southeast corner, and about one third of the north stone barracks.

The state acquired the fort in 1921, and the State Conservation Commission began the present reconstruction work in 1958. The museum was completed in 1962 to display artifacts and information pertaining to the fort and the military forces who served there.

Fort Atkinson can be reached by traveling southwest from Calmar on Highway 24. It is open from May 16 to October 30, daily from 10:00 A.M. to 8:00 P.M. No admission fee.

Bent's Old Fort

LA JUNTA, COLORADO

By THE CLOSE of the American Revolution, New Spain's northern frontier stretched from eastern Texas to the Pacific, effectively barring rich Mexico from intruders. During the late eighteenth and early nineteenth centuries, explorers criss-crossed the Great Plains. After them came fur traders, the vanguard for American expansion into the West. Some pursued the beaver into the northern and central Rockies, others pressed southward into the Southern Plains and lower Rockies, seeking both furs from the Indians and trade with the Mexicans.

When word reached the Mississippi Valley in 1822 that Mexico had thrown off her Spanish shackles, merchants lost no time in testing their information. Several parties set out at once from Missouri for Santa Fe, where they sold their goods at a handsome profit. Soon traders were stopping at Taos and expeditions were pushing on into Mexican territory to trade with Indian tribes there. Their success encouraged others, so that by 1842 the Santa Fe Trail was well established.

Among the first to become interested in trading with the Indians and Mexicans were the brothers Charles and William Bent and Ceran St. Vrain, all sons of prominent St. Louis families. It was the booming Mexican trade of the 1820s that turned their eyes to the Southwest.

After some solid experience on the Upper Missouri fur trade, the three men transferred operations to the Arkansas River, where they built a small stockade near present Pueblo, Colorado. The next year, 1830, they formed the partnership of Bent, St. Vrain and Company. Charles Bent was responsible for arranging credit in St. Louis and purchasing and forwarding goods to New Mexico. William Bent oversaw all Indian trade. St. Vrain, and later Charles Bent, marketed goods in Mexico.

The concept of a great trading establishment on the Arkansas came from Charles Bent. To hold and exploit this territory he knew that they would need a central fort as powerful as those along the Missouri River. The brothers chose a spot on the north bank of the Arkansas, about twelve miles west of the mouth of the Purgatoire River. This placed them just north of the New Mexico boundary, close to the Cheyenne, Arapaho, Ute, Comanche, Kiowa, and Kiowa-Apache, and well within range of roving bands from other tribes. The location also aided the company's business, since trading caravans could go on to New Mexico without leaving the Mountain Branch of the trail.

William Bent started work on the fort sometime in the late 1820s or early 1830s. He built with adobe, both because it was fireproof and because there was little timber available on the

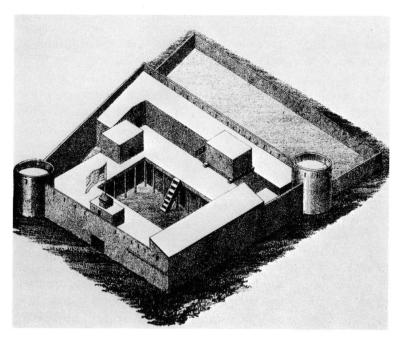

Bent's Old Fort

Overall view of ruins with facilities
(Photo: R. Carrillo)

plains for so large a structure. As more than one hundred Mexican laborers made mud bricks, Americans hauled in timber for roofs and gates. Gradually a building took shape that could be described by a later visitor as one that "exactly fills my idea of an ancient castle." By 1833 the massive impregnable mud fortress stood completed in the midst of an unbroken prairie.

For sixteen years the Bents and St. Vrain managed a private trading empire stretching from Texas into Wyoming, from the Rockies to middle Kansas. By skill and subtleness the Bents achieved greater influence among the Indians than rival traders. Of the numerous Indian tribes trading with the company, the most important were the Southern Cheyenne, upon whose hunting grounds Bent's Fort stood. Bent was "Little White Man" to the Cheyenne and saw that they remained friendly. He required his employees to be fair in bargaining and restricted the use of whiskey. In the mid-1830s Bent married Owl Woman, the daughter of Gray Thunder, one of the most powerful of the Cheyenne. He also encouraged peace among the tribes. The deadliest of enemies could trade at Bent's Fort in an atmosphere of peace.

Bent's Fort was a fairly self-sufficient institution. Employing about sixty persons, it required the services of numerous tradesmen: wheelwrights, carpenters, gunsmiths, and blacksmiths. They were of all nationalities speaking all languages—a perfect Babel of a place. Among the better-known figures of the West employed at the post some time or another were Lucien B. Maxwell, Thomas O. Bogg, Baptiste Charbonneau, and Kit Carson.

When Indian warfare started in 1847, the rich trading days were over. By 1849 the trade which had made Bent's Fort prosper was declining, scores of Indians were dying from cholera, William Bent's three brothers and his first wife had died, and clashes with the Plains Indians along the Santa Fe Trail were becoming more frequent. Bent decided to quit. Removing his family and valuable goods, he set fire to the fort. The amount of charred material found during the archaeological investigation attests to the ferocity to the flames. But Bent wasn't through. Four years later he built a new fort of stone at Big Timbers, about thirty-eight miles downstream. Although fire devastated the interior of the fort, its adobe walls were still standing and subsequently sheltered a stage station. After railroads replaced the stage, the walls served as cattle corrals. They gradually collapsed and disintegrated. By 1915 parts of the old walls were still standing. Elsewhere only mounds outline the fort's dimensions.

Bent's Old Fort National Historic Site was established on March 15, 1963. The National Park Service has conducted a thorough archaeological investigation of the site to determine the actual size and layout of the building. The walls and other features are original.

Bent's Old Fort is eight miles east of La Junta and fifteen miles west of Las Animas on Colorado 194. Open from June to August, daily from 8:00 A.M. to 8:00 P.M.; September to May, daily from 8:00 A.M. to 4:30 P.M. No admission fee.

Fort Bliss Replica

FORT BLISS, TEXAS

IN 1948 on the hundredth anniversary of its founding, Fort Bliss was presented a replica of the original post by the citizens of El Paso. It stands today on the main post of this vast U.S. Military Reservation, home of the Army Air Defense Center, a historical reminder that the United States Army has stood guard at our expanding frontiers throughout the growth of the country.

The first Fort Bliss was established in 1849 in adobe structures on a ranch owned by W. T. Smith. The mission of this post was to protect travelers, pioneers, and settlers from hostile and marauding Indians, to build a wagon road across the territory of the Southwest for emigrants using the southern route to California, and to acquire accurate knowledge of the immense unknown land which was continually being troubled by Indians.

The post was abandoned in 1851, when its

Replica Museum, Fort Bliss, Texas

Replica museum, Fort Bliss, Texas

troops were transferred to other outposts in the Territory of New Mexico. In January 1854 a new post was constructed of adobe buildings on the Magoffin Ranch, about a mile east of the first location. This second post was named Fort Bliss in March 1854. It remained in this location for fourteen years, serving as a base for troops engaged in guarding the area against Indian attacks.

In March 1861 Fort Bliss, along with all other army posts in Texas, was surrendered to the Confederates and then became their head-quarters for the New Mexico-Arizona campaign. It remained in the hands of the Confederates until August 1862, when the 1st California Cavalry drove them well to the east. The Confederates burned most of the buildings and corrals before their retreat, and it was decided not to garrison the post at that time. It was not until the final surrender of all the Confederate armies that Fort Bliss was once more occupied. On October 16, 1865, units of the 5th U.S. Infantry arrived and bivouacked in the area until the buildings could be made habitable.

Over the years the rampaging Rio Grande had eroded away a considerable portion of the post, and the fort was moved to higher ground located a mile farther east. When first occupied in March 1868, that third post was called Camp Concordia and was much larger than the previous posts. It was renamed Fort Bliss in March 1869 and then was abandoned in 1877 because the troopers were needed elsewhere to control the Indians. With the departure of the troops, lawlessness swept the area, and after a complete breakdown of order in El Paso, troops were sent back in 1878. Units of the 9th Cavalry and 15th Infantry marched to El Paso and found the old post in ruins. They were then billeted in rented quarters for two years and drills were held in the public square, now named San Jacinto Plaza.

In 1879 the Army decided to build a permanent post at El Paso, and 135 acres were obtained north of the town on the river. Many of the buildings of that fourth post were of masonry construction, and two of them, built in 1886, are still in use.

When the railroads pushed through to El Paso in 1881, Fort Bliss was directly in the path of their proposed routes. The Santa Fe tracks were laid right through the center of the post. The ever-increasing train traffic created a serious hazard to personnel, and the post was moved to its present site. In 1890 Congress authorized a larger post at a more suitable location and land was purchased at the present site of Fort Bliss. The post has grown with the years, and the entire post today is devoted to guided missiles. The old frontier post which protected the border from Indian and bandit raids has been reconstructed as a historic site. The replica's four adobe buildings are faithful reproductions of those that housed Fort Bliss soldiers from 1854 to 1868. The museum, located in these buildings, depicts the history of Fort Bliss and El Paso through the first one

hundred years of their growth.

The first building contains a chapel honoring Brevet Lieutenant Colonel William Wallace Smith Bliss, for whom the post was named. Also exhibited in the building are pictures of twelve noted travelers who explored the Southwest, and a relief map of the United States shows the routes of the old westward trails. There is a collection of guns showing the evolution of firearms. Leaving the building, a pathway to the right leads through a cactus garden to a simulated cemetery. Displayed near the cemetery is a military ambulance, a Gatling gun, a field blacksmith shop, and a store wagon for hauling supplies. The second building contains a small section of a barracks room, showing a sergeant's full-dress uniform and a corporal's field uniform. Other items displayed include bunks, dragoon pommel holsters, saddlebags, and other military and personal items of equipment used during the 1870 period.

The first section of the third building contains copies of the original drawings of Fort Bliss. The replica is based on these drawings. There are also actual souvenirs retrieved from the battlefields during and after the Civil War—shot and powder bags, minnie balls, powder horns, and uniforms. The second section of this building contains relics from the Moro Rebellion in which the 8th U.S. Cavalry played a prominent role.

The fourth building contains flags, helmets, weapons, and other battlefield souvenirs from World War I and II. Photographs show the development from the beginning of Fort Bliss to its present establishment. The replica grounds have exhibits of military hardware from early defense weapons to modern missiles.

Fort Bliss is approximately two miles north of Interstate 10, Paisano exit. It is open from 9:00 A.M. to 4:45 P.M. daily except New Year's Day, Easter Sunday, Thanksgiving Day, and Christmas Day. No admission fee.

Fort Bridger

CHEYENNE, WYOMING

ONE OF THE most famous of all the early trappers and explorers, Jim Bridger, together with his partner, Louis Vasquez, constructed a trading post in 1842, on a spot near Black's Fork of the Green River in what is now southwestern Wyoming. This original site of Fort Bridger consisted of several crude log buildings and served to supply the emigrants and Indians in the area.

For a decade after 1842 the fort was visited

Fort Bridger (Photo: Wyoming Travel Commission)

Wyoming's first schoolhouse still stands on the grounds of Fort Bridger State Historic Site in southwest Wyoming. (Photo: Wyoming Travel Commission)

48

by Indians, emigrants, gold seekers, adventurers, explorers, and the military expedition under Captain Howard Stansbury in 1849. The fort was not an impressive place in those days; a contemporary account writes of it "built of poles and daubed with mud; it is a shabby concern." Among the early emigrants who stopped at the fort were Mormons fleeing from persecution in the East. Brigham Young and other church leaders wished to have a place where these settlers could rest and obtain supplies. They are reported to have purchased the Fort for six thousand dollars, paid in gold coin to Vasquez in 1855 while Bridger was away. Two years earlier, in 1853, the Mormons had built Fort Supply about twelve miles south of Fort Bridger. The two forts were then used to aid converts to the church as they traveled over the trail to Salt Lake City.

The Mormons occupied the fort for only two years. Friction developed between the newly established Mormon state and the federal government. President Buchanan dispatched United States troops to the area in 1857, precipitating the so-called Mormon War. The Mormons decided to desert and burn their forts because of the advance of an army led by colonel Albert Sidney Johnson and guided by Jim Bridger. In October 1857 when the troops arrived they found the charred remains.

The army began building a permanent fort in the spring of 1858, at first building against the Mormon ruins and later expanding to other locations. By 1859 a total of twenty-nine buildings had been constructed. In honor of his friend and former scout, Colonel Johnston named his post Fort Bridger. The early 1860s were eventful years at Fort Bridger. In addition to military activities the Fort served as a major station for the Pony Express, the Overland Stage Line, and the transcontinental telegraph. Troops provided escort and protection against marauding Indian bands. With the outbreak of

the Civil War, military personnel were ordered East.

For nearly a year the fort was without a garrison until a volunteer militia was organized by available citizens and mountain men. Volunteer regiments from Nevada and California garrisoned the fort from 1862 until 1866. By midsummer of 1866 the last of the volunteers were mustered out and the fort was garrisoned by two companies of regular infantry.

In the late 1860s, detachments from the fort were assigned to escort work crews from the Union Pacific Railway as they made their way west. Many mining expeditions purchased their supplies at Fort Bridger. The fort also served as a vital supply center for troops campaigning in the western portion of the Wyoming Territory.

A period of relative peace settled upon the valley in the 1870s despite the "Indian Wars" taking place in the Northern Plains. The fort was abandoned in 1878 but reactivated on June 28, 1880. Through the 1880s the military erected additional building and barracks, and general improvements were made. But as the frontier became settled and the Indians remained on their reservations, posts like Fort Bridger were no longer needed, and on October 1, 1890, the military permanently abandoned Fort Bridger.

Many early residents remained to make the fort and the valley their home. Fort Bridger was never attacked, nor was it destroyed during or after its military occupations. When it was abandoned, many buildings were sold at auction and some of these buildings are still in use as businesses, private homes, and a school in the Fort Bridger area.

Today, a visitor can see the officers' quarters, the new guardhouse, the sutler's store, the fort cemetery, and a visit to the museum, where many artifacts, documents, etc., are on display, graphically recapitulates the long and colorful history of Fort Bridger.

Fort Bridger Historic Site is located three miles off Interstate 80 near Evanston, Wyoming. It is between Cheyenne and Evanston. It is open seven days a week from April 1 to October 15. No admission fee.

Fort Buford

BUFORD, NORTH DAKOTA

FORT BUFORD was one of the most significant military posts in North Dakota and on the Missouri River. Established on June 13, 1866, near the confluence of the Yellowstone and the Missouri rivers, it was part of a chain of military posts extending from Fort Leavenworth to the Columbia River. It played an important part in quelling Indian difficulties and in settling Indians on reservations.

The first post was to accommodate one company. However, during the first winter, Fort Buford was so harassed by hostile Sioux that it was deemed inadequate. Early in 1867 a rumor was circulated in the newspapers in the East

Officers' quarters, Fort Buford (Photo: State Historical Society of North Dakota)

that the entire garrison was massacred by the Indians and that the commanding officer of the garrison, Captain William O. Rankin, had killed his wife to prevent her from falling into the hands of the Indians. In the following summer, a new five-company post was constructed mainly of adobe and surrounded by a stockade twelve feet in height, which was subsequently removed in 1871. Materials salvaged from nearby Fort Union trading post was used in the construction of the post. The reservation was enlarged in 1868 to include an area thirty miles square.

Located in the heart of hostile Indian territory, Fort Buford was under constant siege during its early days. While war parties never attacked the fort itself they made raids on woodcutting, haying, and hunting parties so frequently that it was impossible for men to leave the fort without an armed escort. In August 1868 two to three hundred Indians divided into two bands, hid in the ravines of the Badlands about Fort Buford, and raided a herd of about 280 cattle escorted by a mounted guard of twenty men. They succeeded in driving off all but about forty of the herd. The Indians also attacked mail parties operating between the post and Fort Stevenson.

At first, Fort Buford largely served as police to the region. It restrained traders from illegally trading whiskey and ammunition to the Indians in exchange for fur pelts. The post also distributed annuities from the government to the various Indian tribes living in the vicinity.

Fort Buford achieved its greatest importance during the Indian Wars of the late 1870s and early 1880s. Following the Custer disaster on the Little Bighorn, the government conducted a relentless campaign against the Sioux and Cheyenne in eastern Montana and in Wyoming. Many of the men and supplies were shipped up the Missouri and Yellowstone to support this campaign. The fort served as a depot of supplies for these operations. Numerous detachments were also supplied for the escort of trains passing up both of these rivers. After the surrender of Chief Joseph and his followers at Bearpaw Mountain in Montana in 1877, the Nez Perce prisoners-of-war were brought to Buford from which they were shipped to Bismarck and on to Fort Leavenworth in Kansas. Many warriors of Sitting Bull who had escaped into Canada and then were driven back into Montana by severe cold and scarcity of game, surrendered at Fort Buford early in 1881. Sitting Bull and 187 men, women, and children surrendered on July 19 and later transported to Fort Yates. Sitting Bull was sent from there to Fort Randall and kept in custody for some time. He spent his final days at the Standing Rock Agency where he met his death in 1890 at the hands of the Indian police.

Because of the post's isolation, the officers and men of Fort Buford led a very monotonous existence broken only by the visits of steamboats during the summer months and the irregular arrival of the mail. Complaints of drunkenness were frequent among the officers, who used all holidays as occasions to go on a spree. There was a library of 366 volumes of "principally light reading," and three theatrical groups and the regimental band.

Much of the time was devoted to policing the border to prevent Indians from crossing from Canada into the United States to make raids on the American reservations and keeping those from the United States from crossing into Canada. When the Great Northern Railway was built through their region, Fort Buford provided troops to guard the construction workers.

As white settlers poured into the region in the late 1880s and early 1890s, it became increasingly evident that the post was no longer needed. In August 1895 Brigadier General

John R. Brooks reported that the buildings were in a dilapidated condition and no longer of value as a military post and recommended its abandonment. It was officially abandoned on October 1, 1895, and the four companies stationed there were transferred to Fort Assinboine, Montana. The buildings were subsequently sold at public auction.

It was established in 1924 as a state historic site. Among the surviving structures at the fort are the officers' quarters, which now serves as a museum, and the powder magazine. The original stone powder magazine still stands.

Fort Buford is located south of Williston on Highway 2. There are accommodations for trailer camping. No admission days, hours or admission fees specified.

Campus Martius

MARIETTA, OHIO

Forty-seven New England men landed at the mouth of the Muskingum River on April 7, 1788. They were members and employees of the Ohio Company of Associates which had secured from the Continental Congress a 1.8-million-acre land grant. The pioneers set to work laying out the town of Marietta, named for the French Queen Marie Antoinette. This was the first permanent settlement and seat of government to be established in the Northwest Territory after the passage of the Ordinance of 1787. During the following months, families of the pioneers and other adventurers floated down the Ohio River on flatboats to settle in the log houses constructed for them in the village.

The cold fear of Indian attack haunted the settlers perpetually. Fort Harmar, across the Muskingum, could afford little protection to Marietta in event of a raid. The leaders of the settlement decided to plan a fortified village to be built on a bluff above the Muskingum River three quarters of a mile from its junction with the Ohio.

Work was begun on the fort early in the summer of 1788, and on July 2 it was named Campus Martius, after its counterpart in ancient Rome, the words literally meaning "field of Mars."

The four walls of the fort measured 180 feet each on the outside. At each corner a two-story blockhouse crowned with a sentry tower was erected. Homes of individual settlers were built on plots of land leased by the company between the blockhouses. The walls of the houses were constructed with four-inch-thick planks to form the curtain of the fort. A course of pointed, thorny sticks and log palisade completed the outer defenses.

This unique combination of dwelling and defensive perimeters lasted until peace was ensured on the frontier following the Treaty of Greene Ville, when the fort was abandoned. The superintendent of the Ohio Company, Rufus Putnam, remained in his Campus Martius home while other residents moved to new houses. Using timber from one of the abandoned blockhouses, Putnam's carpenters added four new rooms across the front of the original structure and here Putnam lived until his death in 1824.

Today, the Putnam House, the last remainder of a fortified village, can be seen—the oldest known house preserved in Ohio. It is enclosed on its original foundation in one wing of Campus Martius Museum. The exhibits show a complete picture of domestic life in Campus Martius and in the Old Northwest and Ohio before the Civil War.

You can reach Campus Martius in Marietta at the corner of Washington and Second streets.

Open Monday to Saturday from 10:00 A.M. to 6:00 P.M.; Sunday from 1:00 to 6:00 P.M. Adults $1.00. Children under thirteen free if accompanied by parent.

Fort Casey

COUPEVILLE, WASHINGTON

FORT CASEY is a good example of a system of coastal forts which, during a period of about twenty years (1898-1918), was a primary line of defense for the United States. Along with batteries at Fort Worden and Fort Flagler, its guns guarded the entrance to Admiralty Inlet, the key point in a fortification system designed to prevent a hostile fleet from reaching

Fort Casey (Photo: Interpretive Div., Washington State Parks)

Disappearing gun, Fort Casey

such prime targets as the Bremerton Navy Yard and the cities of Seattle, Tacoma, Olympia, and Everett.

A small detail arrived on the reservation shortly after the completion of the gun emplacements. The first garrison numbered thirty men under the command of Lieutenant A. D. Putnam. Fort Casey was officially activated in 1900, and although its guns were never fired in anger, it remained an integral part of the United States defense until 1921 when it was placed on caretaker status.

Mounting of the guns in the completed batteries was accomplished by January 26, 1900. The first test firing was performed on September 11, 1901. A characteristic feature of early coastal forts such as Fort Casey was the large guns mounted on disappearing carriages. Guns of this type were withdrawn behind a thick concrete parapet after each round was fired. The main armament of Fort Casey consisted of seven ten-inch disappearing carriage guns. In addition there were six six-inch disappearing carriage rifles, two five-inch rapid-fire guns on balanced pillar mounts, four three-inch rapid-fire guns on pedestal mounts, and sixteen twelve-inch mortars.

The new reservation and batteries on Admiralty Head were officially named Fort Casey by the Army's Adjutant General's Office in 1899, in honor of Brigadier General Thomas Lincoln Casey, last Chief of Engineers.

In common with other coast defense posts in the west, Fort Casey was the scene of increased military during World War I. After World War I the fort contained only a small force, at one time amounting to a single platoon under a sergeant. It was kept at the post to guard and maintain the property. Military activity during World War II consisted largely of training and routine garrison duty. A number of the remaining outmoded coast defense guns were salvaged for scrap in 1942 and 1943. All of the guns were removed prior to its abandonment in 1950.

The Washington State Parks and Recreation Commission acquired Fort Casey in 1956 and initiated an interpretive program designed to provide the public with a complete story of the Coast Artillery. An interpretive center on coastal defense forts has been installed in the old Admiralty Point Lighthouse, and two three-inch rapid-fire guns are on display at Battery Trevor to the southeast of the central parking lot.

Fort Casey Coastal Defense Heritage Site is located on Whidbey Island, three miles south of Coupeville. It is open to the public April 1 to October 15, from 6:30 A.M. to 10:00 P.M.; October 16 to March 31, from 8:00 A.M. to 5:00 P.M. No admission fee.

Fort Caspar

CASPAR, WYOMING

PAST THE LOG BUILDINGS of Fort Caspar the mountain men of the fur trade, the covered-wagon emigrants on their way to Oregon territory, the California-bound "Forty-niners," Mormon pioneers headed for Salt Lake Valley, passengers on the first transcontinental stagecoaches, Indians on their way to the treaty grounds near Fort Laramie, and riders of the Pony Express, all went their designated ways. This was the traffic the fort saw since it was built in 1857 for United States Cavalry to guard the important transcontinental Oregon Trail route. Its mounted troops also protected the Platte Bridge, the main crossing of the Oregon Trail over the North Platte River, during those covered-wagon days.

Originally named Platte Bridge Station, the fort was renamed for Lieutenant Caspar Collins, who, with a detachment of twenty-five men, faced two thousand Indians and was killed. His death was described: "Shortly after leaving the military outpost, Collins' command was attacked by Indians. Severely wounded Collins ordered his men to retreat to the Platte River Bridge. He managed to fight his way clear when one of the wounded soldiers cried out for help. He whirled his horse around to go to the help of the man. Still mounted he managed to get the wounded man off the ground and on his knees. His horse suddenly reared and the wounded man fell back to the ground. The horse stampeded and Collins was carried into the midst of the Indians. He fought, was surrounded and killed."

Fort Caspar stood witness to the westward march of history in the mid-1800's. (Photo: Wyoming Travel Commission)

The present fort on the western edge of Caspar is a replica of the original. It was rebuilt on its original site, faithfully replicated from contemporary records. The several buildings now house Indian artifacts and pioneer relics.

U.S. Highway 26 and State Highway 220 lead to Caspar. The fort is open from May 15 to September 15, daily from 10:00 A.M. to 9:00 P.M. No admission charge is listed.

Fort Clatsop

ASTORIA, OREGON

THE LEWIS AND CLARK EXPEDITION—the first journey across the North American continent between the Spanish possessions on the south and British Canada to the north—was an event of major importance in the history of the United States. The expedition gave the first detailed knowledge of the American Northwest and started the procession—first of trappers, then of settlers—that was a factor in making Oregon American rather than British.

On May 14, 1804, the expedition started from the mouth of the Missouri River near St. Louis in one fifty-five foot keelboat and two smaller open boats called pirogues. The 1,600-mile ascent of the Missouri to the Mandan villages in what is now North Dakota was a tedious voyage of more than five months. Only one life was lost and that due to natural causes.

Fort Mandan was built and occupied during the long, hard winter of 1804–5. On April 7, 1805, the party of thirty men plus the Charbonneau family of three (half-breed interpreter for the party), left Fort Mandan in two pirogues and six canoes to explore the unknown. The continued ascent of the upper Missouri presented many challenges—a long portage around the Great Falls, mountain crossings, etc. On November 15, after some 600 miles of water travel, they had their first full view of the ocean from near present-day McGowan, Washington.

Their camp on the north shore of the estuary was exposed to the ocean gales, and the hunting was poor. Learning from visiting Clatsop Indians that elk were more plentiful on the south side of the Columbia, they resolved on a location two miles up a small stream, now called the Lewis and Clark River, flowing into a bay of the Columbia now known as Young's Bay. The site was chosen because it was near good hunting in the lowlands where elk wintered in large numbers. It was thirty feet above high-water mark, suitable timber for construction was at hand, and freshwater springs were close-by. The main village of the friendly Clatsop Indians was about eight miles distant. On December 8 construction was begun and the party was under shelter on Christmas Day, 1805. The structure was named Fort Clatsop, after the local Indian tribe.

The Clatsop, Chinook, Cathlamet, and Tillamook were the most frequent visitors at Fort Clatsop. They lived in the surrounding areas. They came to visit and to trade, bringing fish, roots, furs, and handcrafted articles. There were practically no hostile incidents. Clark described them as close bargainers. They were fond of smoking but did not drink whiskey.

At the fort strict military routine was observed. A sentinel was constantly posted, and at sundown the fort was cleared of visitors and the gates shut for the night. There was never much food in reserve and hunting for meat was ever-important. Cutting firewood in the dripping rain forests was a continuous task. A trail to the seacoast was established for the use of hunters and salt-makers. Both Lewis and Clark made copious notes on the trees, plants, fish, and wildlife of the vicinity. Many such descriptions were the first identification of important flora and fauna of the Pacific Northwest. Clark, the cartographer of the party, spent most of his time drawing maps of the country through which they had come.

As spring approached, the elk took to the hills, and it became increasingly difficult for the hunters to keep the camp supplied with meat. The men were restless and anxious to begin the return trip. Fort Clatsop, with its furnishings, was presented to Comowool, the Clatsop chief, as a mark of appreciation for his cooperation and friendliness. On March 23, 1806, the expedition embarked in canoes for the trip up the Columbia River and thence to the United States and home.

Nothing of the original fort has survived.

Through the cooperative efforts of citizens and organizations of Clatsop County, a replica was built in 1955 on the occasion of the Lewis and Clark sesquicentennial celebration. It faithfully follows the floor plan dimensions as drawn by Captain Clark on the elk hide cover of his field book. Since no contemporary drawing of the building is known to exist, the general appearance is largely conjectural, but based on knowledge of similar structures of that period.

The site of Fort Clatsop was preserved by the Oregon Historical Society and later donated to the people of the United States. It is now a National Monument under the National Park Service. It is located four and one half miles southwest of Astoria. U.S. 101 passes just north of the area. It is open June 15 to Labor Day, daily from 8:00 A.M. to 8:00 P.M.; Labor Day to June 14, daily from 8:00 A.M. to 5:00 P.M. No admission fee. A museum interpreting the story of the Lewis and Clark Expedition is part of the site.

Fort Columbia

CHINOOK POINT, WASHINGTON

AT THE MOUTH of the Columbia River near Chinook, between the towns of Ilwaco and Megler in Pacific County, stands Fort Columbia, one of the three fortifications built by the government to guard the mouth of the Columbia River. In 1845 it appeared that the United States and Great Britain might go to war over the Oregon Country, which had been open to settlement by the citizens of both countries under a joint occupation agreement. In order to prepare for any eventuality, the British government sent Lieutenants Henry J. Warre and Mervyn Vavasour from Canada to make a military reconnaissance of the disputed area.

The two officers examined the mouth of the Columbia during the winter of 1845–46 to plan for its defense. Chinook Point was among the locations they believed might serve "for temporary purposes." As matters developed, the international controversy was settled by treaty during 1846, and the Oregon Country south of the 49th parallel became United States territory.

Soon after the close of the Mexican War the United States took measures for the defense of the vast new territory acquired in the West as a result of that conflict and of the Oregon Treaty of 1846. Engineers were sent to the Pacific

Administration building, Fort Columbia

Coast to select sites for fortifications. One result of such surveys was the setting aside in 1852 of land at the mouth of the Columbia River for military purposes. The lack of appropriations caused delay after delay. Probably the fear of Confederate cruisers, particularly the *Alabama*, induced Congress to appropriate $100,000 and the War Department to hurry the building of Fort Columbia. Legal delays in acquiring the land caused an indefinite postponement of the work. By the time the title was settled the War Department evidently changed its mind about the urgency of the need for fortifications on Chinook Point.

It wasn't until 1896 that construction of Fort Columbia was begun. During the construction period from 1896 to about 1904 the War Department also modernized and strengthened Fort Stevens and Fort Canby. When all three were completed, the fortifications at the entrance to the Columbia had assumed approximately the form they maintained until World War II.

The first project authorized for the Fort was an eight-inch gun battery. A bit later a concrete mining casemate was added. Excavation for a battery of two guns was started during March 1897 on the slope of Scarboro Hill at an elevation of one hundred feet. By April 1898 two eight-inch rifles with disappearing carriages were mounted. A third emplacement for one eight-inch rifle was completed in May 1898 and armed in June of the same year. This line of three eight-inch rifles was later known as "Battery Ord" in honor of Lieutenant Jules G. Ord, killed in action at San Juan Hill, Cuba, on July 1, 1898.

By July 16, 1898, the work contemplated in the first allotments for Chinook Point had been finished, and on that day the engineers turned the emplacements, the ordinance, and the concrete mining casemate over to the commander of Fort Stevens. Eleven men of that post were

sent on July 15, 1898, to guard the new fortifications. They were the first troops to staff the future Fort Columbia. Additional fortifications were built during the next two years. Battery Crenshaw, consisting of three three-inch rapid-fire guns were constructed low on the south side. Almost adjoining it to the northwest was Battery Murphy, completed in June 1900. It contained two six-inch disappearing-type guns. The principal buildings needed to house a garrison were erected on the hillside above the batteries. Completed by the end of 1902 these were a large barracks, one single and one double set of officers' quarters, one double set of noncommissioned officers' quarters, a hospital, a guardhouse, and an administration building. All were of frame construction.

Until World War I the garrison at Fort Columbia generally numbered about four officers and one hundred men. The outbreak of war in 1917 saw increased activity and its facilities were renovated and modernized. In 1941 construction of barracks to house additional troops was begun and the old buildings were rehabilitated. The declaration of war in 1941 greatly increased military activity in the Harbor Defenses of the Columbia. The minefields at the river entrance were put in a state of readiness. Improvements made in the mine control equipment at Fort Columbia during the war included a new gasproof and splinterproof casemate and modern mine control communication facilities. Under the impetus of the war a new two six-inch battery of long-range, rapid-fire rifles mounted on barbette carriages was planned. Begun in 1942 this battery was never completed. A tract of forty acres was also added to the Fort Columbia reservation.

Throughout its history Fort Columbia never fired a shot "in anger." The closest it came to it was on the night of June 21, 1942, when a Japanese submarine fired nine shells which fell harmlessly at Fort Stevens. This event has been termed the only bombardment by a foreign craft of a fortification within the continental limits of the United States since the War of 1812. Observers at Fort Columbia saw the flashes, estimated the range of the enemy, but did not return the fire. The submarine was out of range of the fort's guns.

On March 28, 1947, the three forts of the Harbor Defenses of the Columbia were listed as surplus. Fort Columbia was stripped of its armament. The Washington State Parks and Recreation Commission applied for the proper-ty for historical monument purposes, and in June of 1951 the old military post became Fort Columbia Historical State Park. The old gun emplacements, some of the buildings, and an Interpretive Center featuring exploration are open to visitors.

Fort Columbia is at the mouth of the Columbia River near Chinook. State Road 12A South leads to Chinook. It is open from 6:30 A.M. to 10:00 P.M. seven days a week from April 1st to October 15 and from 8:00 A.M. to 5:00 P.M. during the off season. No admission fee.

Fort Davis

FORT DAVIS, TEXAS

THE MEXICAN WAR of 1846–48 added to the United States a vast territory comprising the present states of New Mexico, Arizona, and California. Texas had joined the Union on the eve of the war. Interest in the new lands quickened when word of the discovery of gold in California burst upon the nation in 1849. Intent upon avoiding the winter snows and rugged mountains of the central routes to the gold fields, thousands of immigrants made their way over the southern transcontinental trails. A vital segment of this route was the newly opened San Antonio–El Paso road. Intersecting the El Paso road were the raiding trails of Indians who swept southward on plundering expeditions that had long devastated the isolated villages and haciendas of northern Mexico.

West of the Davis Mountains the Mescalero Apaches from New Mexico crossed the road in their forays. East of the mountains, the "Great Comanche War Trail" crossed the road at Comanche Springs, later the site of Fort Stockton. Both tribes raided travelers on the El Paso Road. By 1854 Indian depredations had grown to such horrendous proportions that the military authorities in San Antonio deemed it essential to build a fort in West Texas.

In October 1854 the commander of the Department of Texas, Major General Persifor F. Smith, personally selected the site, a pleasant box canyon near Limpia Creek in scenic Davis Mountains. The new post would be named Fort Davis in honor of Secretary of War Jefferson Davis. Six companies of the 8th U.S. Infantry marched westward to build and garrison Fort Davis. Their commander, Lieutenant Colonel Washington Seawell, disliked the site, for Indians could and did come very near without discovery. Dutifully, however, he placed the fort where ordered. It was a shabby collection of more than sixty pine-slab structures scattered irregularly up the canyon. They were built not to last, for Colonel Seawell planned someday to erect a fine new stone fort on the open plane at the mouth of the canyon. He spent most of the next seven years as commander of Fort Davis but never realized his ambition for a permanent post of stone buildings. By 1856 he had housed the enlisted men in six

Fort Davis

Fort Davis

stone barracks laid out in a line across the mouth of the canyon. The officers continued to live in rotting log huts and the supplies deteriorated in rickety warehouses roofed with canvas or thatched with grass.

The men spent most of their time, afoot or mounted on mules, escorting mail and freight trains, pursuing but rarely catching raiders who had attacked travelers or a mail station, and covering their sector with patrols that rarely come to grips with Apaches or Comanches. On occasion, cavalry expeditions used the fort as a base for concerted operations against the Indians. New forts—Hudson, Lancaster, Stockton, and Quitman—were built to aid in the task of guarding the El Paso road. From 1857 to 1860 the feasibility of using camels for military purposes on the western deserts was tested, with encouraging results, at and near Fort Davis. By the close of the decade, however, little real progress had been made in solving the Indian problem.

The Civil War destroyed the frontier defense system of West Texas. With the secession of Texas from the Union early in 1861, the department commander, Brigadier General David E. Twiggs, ordered Fort Davis and her sister posts abandoned. They were occupied by Confederate troops in June 1861. In August of 1861 the Mays Massacre occurred. Lieutenant Reuben E. Mays and fourteen men who were pursuing Apaches who had stolen horses and cattle from the fort were ambushed by eighty to one hundred warriors and wiped out.

After the war the Indian menace to West Texas had ended and cattlemen began to drive their herds westward to pasture them on the rich grasses of the region. The railroads bypassed the fort, and it became increasingly an unnecessary expense. In June 1891 Fort Davis was ordered abandoned.

Of more than fifty adobe or stone buildings that constituted Fort Davis when it was abandoned, visitors can inspect eighteen residences on officers' row, two sets of troop barracks, warehouses, and the hospital. Marked sites of the remaining buildings—stone foundations in most instances—can also be viewed. Fort Davis is a National Historic Site.

Fort Davis is located on the northern edge of the town of Fort Davis, Texas, and can be reached from U.S. 290 on the north and 90 on the south by paved Texas 17 and 118. It is open from 8:00 A.M. to 5:00 P.M. daily. No admission fee.

Fort Dodge

DODGE CITY, KANSAS

FORT DODGE, named for Major General Grenville M. Dodge, was established in 1864 as a supply depot and base of operations against warring Plains Tribes. In its day, the fort was one of the most important military establishments on the western frontier. It was located on the north bank of the Arkansas River, a short distance southeast of present-day Dodge City, on the site of the "Caches," which had been a favorite camping ground for freighters and hunters from the time of the opening of the Santa Fe Trail.

There is some question about the true origin of the fort's name: Some say that a fort was located in this place in 1835 built by Colonel Henry I. Dodge. Records show that Colonel Dodge did erect some sort of a fort in the locality, but the reports of the U.S. War De-

Building once used as officers' quarters, Fort Dodge (Photo: Kansas State Historical Society, Topeka)

partment state that Fort Dodge was established by General Grenville M. Dodge in 1864 and that the site was selected by Colonel Ford, of the Second Colorado Cavalry.

Like the other forts bordering the Santa Fe Trail, the most important highway to the West, Fort Dodge was active in protecting traders and travellers from Indians and highwaymen throughout its distinguished history. It boasted one of the finest garrisons in the country, and, at one time, General George A. Custer was the commanding officer of the post.

The first buildings were of adobe, but in 1867 several new structures were erected at considerable cost. After trade and settlers located safely to the west and troubles with the warring Plains Tribes diminished, the usefulness of Fort Dodge declined and the fort was finally abandoned in 1882. The enclosure of the fort was used as a cattle corral and the buildings fell into decay. In 1889 the reserva-

tion was donated to the state to be used as a site for a soldier's home.

Today, two of the original adobe structures still stand, although they have been veneered with stone: the commandant's building, now the superintendent's home, and another building, now used as the administrator's building. These were among those built in 1867. There are five stone buildings which cannot be definitely dated but remain from the days of military occupancy: the old fort hospital, now "Pershing Barracks," housing residents of the soldier's home; the present library building, presumed to be the old fort commissary; and three small stone cottages. The old jail has been moved to "Boot Hill" in Dodge City.

Fort Dodge, now a state soldiers' home, is located four miles southeast of Dodge City on U.S. 154. Admission by permission. No admission fee.

Fort Fetterman

DOUGLAS, WYOMING

INDIAN ATTACKS upon civilians traveling north along the Bozeman Trail to Montana made that trail an uncertain and precarious travel route. Wagon trains were advancing across the frontier along the line of the Union Pacific Railroad, and it was decided that building a fort at a strategic spot along the route would furnish the Army a badly needed supply point to be used in operating against the Indians.

In the words of a contemporary, Fort Fetterman "is situated on a plateau (containing nearly one square mile) above the valley of the Platte, being neither so low as to be seriously affected by the rains or snow; nor so high and unprotected as to suffer from the (prevailing

westerly) winter winds. It is on the south side of the Platte and east of the LaPrele and is within two hundred yards of either water course. There are three fords of the Platte within the vicinity which have hitherto been used by the Indians; one is about half a mile above and the others within three or four miles below."

The fort was named for Brevet Lieutenant Colonel William J. Fetterman, infantry captain who, with his whole command, was killed in the Fetterman Massacre at Fort Phil Kearny, in northern Wyoming, on December 21, 1866. Construction of the fort was started early in July 1867. Major William McEnery Dye, with companies A, C, H, and I of the Fourth Infantry, was assigned to build the post. Major

Fort Fetterman (Photo: Wyoming State Archives and Historical Department)

Dye was the contemporary whose view of the new fort site was so optimistic. This view was not shared by Brigadier General H. W. Wessels, who became commanding officer at Fort Fetterman in November. According to his report to the Department of the Platte, "officers and men were found under canvas exposed on a bleak plain to violent and almost constant gales and very uncomfortable." The garrison managed to get through the winter and the fort continued to grow and develop until, by 1870, it was well established and destined to play a conspicuous part in the Indian wars for the next few years. Jim Bridger, Wild Bill Hickock, Calamity Jane, and "Buffalo Bill" Cody were among the colorful personalities whose activities and travels took them frequently to Fort Fetterman.

In accordance with the Treaty of 1868, Forts Reno, Phil Kearny, and C. F. Smith, along the Bozeman Trail, were abandoned. Fort Fetterman alone remained on the fringe of the disputed area. As an outpost of civilization of the western frontier, the fort represented protection and haven to travelers.

Fort Fetterman was always considered a hardship post by the officers and men who were stationed there. On May 18, 1874, Captain F. Van Vliet, of Company C, 3rd Cavalry, felt so strongly about the hardships on his men that he wrote to the Adjutant General requesting that his company be transferred because there was "no opportunity for procuring fresh vegetables, and gardens are a failure. There is no female society for enlisted men. The enlisted men of the Company are leaving very much dissatisfied, as they look upon being held so long at this post as an unmerited punishment. Whenever men get to the Railroad there are some desertions caused by the dread of returning to his post."

Desertions were common, and the post frequently lacked adequate supplies and equipment. Supplies had to be hauled from Fort Laramie to the southeast, or from Medicine Bow Station on the Union Pacific Railroad. Luxuries were scarce and pleasures few. However, the soldiers found some diversion from the routine of garrison life at a nearby establishment known as the "Hog Ranch."

During the mid-1870s, Fort Fetterman reached the peak of its importance when it became the jumping-off place for several major military expeditions. It was the base for three of General George Crook's Powder River Expeditions and Colonel Ronald Mackenzie's campaign against Dull Knife and the Cheyenne Indians. These campaigns contributed to the end of the resistance by the Plains Indians, who, shortly after, were confined to reservations. With the passing of the Indians from the scene, the fort, like many others in similar circumstance, had outlived its usefulness.

When the military abandoned the fort in 1882, life continued for several more years. A community grew up at the post, and, after 1882, it was an outfitting point for area ranchers, and for wagon trains. In 1886 these activities were shifted to the town of Douglas, founded a short distance to the south. The old fort, in a state of decay, lost out as a town and declined rapidly. Most of the buildings were sold, dismantled, or moved to other locations.

Fort Fetterman today is preserved by the State of Wyoming as a State Historic Site and Museum. A restored officers' quarters and an ordinance warehouse are the only original buildings left at the site. A museum of the fort's history is housed in the officers' quarters. Each year, during the annual Fort Fetterman Days celebration held the third weekend in July, volunteers dressed in authentic period costumes portray the life of the soldier and citizen of the 1870s.

Fort Fetterman is northwest of Douglas.

Highway I-25, the road from Cheyenne to Casper, has a turnoff to Orpha, that leads to the fort. The fort is open seven days a week from May 1 to September 30. No admission fee.

Fort Garland

FORT GARLAND, COLORADO

ORT GARLAND was established by the War Department in 1858, in what was then the Territory of New Mexico and is now Costilla County, Colorado. It replaced Fort Massachusetts, which had been built in 1852 six miles north. Fort Massachusetts was abandoned because it was so situated that hostiles could fire from the hills down into the fort. The new fort was named after the commander of the Department of New Mexico, Colonel Brevet Brigadier General John Garland. For twenty-five years it was a garrison for troops protecting settlers.

Fort Garland was build by the men of Company E, U.S. Mounted Riflemen, and Company A, 3rd U.S. Infantry, under Captain Thomas Duncan of the Mounted Riflemen. It was planned for two companies of about one hundred men and seven officers.

The buildings were made of adobe, their interiors plastered with mud and whitewashed with lime. Roofs were of sod. There were board floors, and open fireplaces, with stoves for additional heat. Water was obtained from Ute Creek, by means of an *acequia* (Spanish for canal or trench) flowing around the parade ground. Sanitary provisions, even for the hospital, were primitive. When the fort was built, the nearest railroad was 950 miles away, and six weeks were required for mail to go from Fort Garland to Washington.

The post buildings formed a parallelogram around the parade ground. On the north side were the officers' quarters, on the east the cavalry barracks, on the west the infantry barracks, and on the south were two long buildings which served as offices and storerooms and housed the guardroom and adjutant's headquarters.

The two units that constructed the buildings remained at the fort until July 1860, when they were relieved by Companies A, F, and H of the 10th Infantry under Major E. R. S. Canby. The 10th U.S. Infantry had formed part of the expedition to Utah under Colonel A. S. Johnston during the "Mormon War," and Canby's detachment marched the 640 miles to Fort Garland from Camp Floyd, Utah Territory, near Salt Lake City.

In August 1860 Company A of the 10th U.S. Infantry left the post to take part in an expedition against the Navajo Indians of New Mexico. In 1861, after the Civil War began, the other units of the 10th were sent to New Mexico, under Canby, except for a small detachment left at Fort Garland. Two companies of Colorado Volunteers arrived in December 1861 and were mustered into federal service at Fort Garland as Companies A and B, 2nd Colorado Infantry Volunteers. They departed in 1862 to join Canby's forces in New Mexico. During the rest of the Civil War, the garrison consisted largely of volunteers, the regulars all on duty with the Union forces in the main theaters of war.

At the end of the war, some of the volunteer units were retained in federal service pending return of the regular troops to man the western

Officers Row, Fort Garland (Photo: Library, State Historical Society of Colorado)

frontier outposts. One of these units was a regiment of New Mexico Volunteers commanded by Kit Carson, a colonel and brevet brigadier general. He was ordered to Fort Garland with a detachment of his regiment in 1866.

The Ute Indians were warlike and the settlers in the San Luis Valley were apprehensive. Kit Carson knew the Ute and had their confidence. He averted open war and saved the settlements. General William Tecumseh Sherman paid tribute to Carson's influence with the Ute when he visited the fort on an inspection trip in 1866 and conferred with Ouray, the Ute chief. In the summer of 1867 the Volunteers were mustered out and Carson, his wife and six children moved to Boggsville, near present Las Animas, Colorado. There, in 1868, he became ill and was taken to nearby Fort Lyons, where he died, at the age of fifty-eight.

Fort Garland was again garrisoned by regulars, infantry, and cavalry at various times. During the critical period following the Meeker Massacre at White River Agency in 1879 and the subsequent removal of the Uncom-

pahgre Ute to Utah in 1881, the garrison was considerably increased. At one time, during this period, there were fifteen hundred men camping in the fort area. With the Ute on reservations, and other Indians peaceful, Fort Garland was no longer needed, and on November 30, 1883 it was abandoned as an active post of the U.S. Army. Its last garrison was Company A, 22nd Infantry, Captain Javan B. Irvine commanding.

In 1928 a group of public-spirited citizens of the San Luis Valley bought the property to preserve the fort, and in 1945 they presented it to the State Historical Society of Colorado, which restored the existing buildings as nearly as possible the way they were when the fort was in use.

Fort Garland is twenty-five miles east of Alamosa and thirty-three miles from the northern boundary of New Mexico via U.S. 6 and State 91. It is open daily from 9:00 A.M. to 5:00 P.M. from June 1 to October 15. No admission fee.

Fort Gibson

FORT GIBSON, OKLAHOMA

THE OLDEST military post in Oklahoma, Fort Gibson was built in 1824 on the left bank of the Neosho (Grand) River three miles above its confluence with the Arkansas River across from Muskogee. It was originally built to serve as protection for the immigrant Indians who had been removed from their homes in the southeast to what is now Oklahoma. This was the Osage tribe. The immediate cause for the establishment of the post was the massacre of a party of white trappers on the Blue River in November 1823. It was intended to restrain the Osage Indians to end their fighting with the

Cherokee and to protect white hunters in the area. Gibson was the Army's first move westward from Fort Smith, and for a time it was the farthest western outpost in the United States.

Its location was highly strategic, and for a half century it had an important civilizing influence on the entire Southwest. It was, according to one historian, "a frequent conference site, outfitting point for military and exploratory expeditions, and rendezvous for army officers, government commissioners, writers, artists, missionaries, traders and adventurers." It was named for Colonel George Gibson, com-

Parade grounds, Fort Gibson (Photo: Fred W. Marvel)

missary general of the U.S. Army.

The fort was evacuated in May 1836, when the entire garrison was ordered to the Texas border. It was reoccupied in January 1837, when the rumored Mexican invasion of the Republic of Texas failed to materialize. For a time Gibson had the reputation for being "the graveyard of the Army." This situation was eased somewhat with start of work on the permanent stone buildings in 1845. One barracks, the commissary, the commandant's house (now a private residence), and several other structures were erected. Drinking was a major problem at the fort. One post commander tackled the problem by standing the offender on the head of a barrel, an empty bottle in each hand, a board around his neck reading, "Whiskey Seller."

Ordered abandoned on June 8, 1857, and substantially evacuated by June 22, 1857, the last troops left the post in September. The land and buildings were turned over to the Cherokee Nation. It was occupied by Confederate troops early in the Civil War and retaken by a force of Union troops, including Cherokee volunteers, on April 5, 1863. The buildings of the old post were repaired, new buildings were erected, and extensive earthworks constructed. It was broken up as a military post in 1871 and finally abandoned in 1890. Many of the post buildings were torn down and others were converted into private dwellings.

Many famous people lived at or visited Fort Gibson. Three presidents of independent republics—Jefferson Davis of the Confederacy, Zachary Taylor of the United States and Sam Houston of the Republic of Texas—were familiar faces at the fort. Other outstanding people to visit the fort were Washington Irving, George Catlin, and Robert E. Lee.

The old stockade, which was abandoned in 1890, has been realistically restored. It contains six rooms of artifacts associated with the fort. Many of the original buildings still stand but are in private hands and may not be visited.

Fort Gibson Stockade is located one mile north of Fort Gibson, Oklahoma, on Highway 80. It is open on Tuesday and Sunday from 1:00 to 5:00 P.M.; from Wednesday through Saturday from 9:00 A.M. to 5:00 P.M. (open until 7:00 P.M. on above days during daylight savings time); closed on Monday. No admission fee.

Fort Harker

KANOPOLIS, KANSAS

IN THE MID-1860s, marauding bands of Indians caused consternation and fear to the settlers of Kansas. They demanded and got one of the best means of defense against random Indian attacks, a chain of forts designed to protect settlements and travel routes. Fort Harker was one of the chain of forts built to guard the settlers of the region and the route of the Butterfield Overland Despatch and the Smoky Hill shortcut to Denver, "which passed through the most Indian-infested region in Kansas." Others in that same link were Fort Wallace and Fort Hays. Fort Harker was also a link in the military trail from Fort Leavenworth, Fort Riley through Fort Zarah, Fort Larned, and Fort Dodge. With the coming of the U. P. Railroad, many using the Santa Fe Trail took the railroad to Fort Harker and then followed the Harker-Zarah trail to the Santa Fe Trail and on to New Mexico.

Historic wagons, Fort Harker

Fort Harker Guardhouse (now museum) (Photo: Kansas State Historical Society, Topeka)

During its active career of nine years, Fort Harker proved to be a bulwark of defense against hostile Indians. It was one of the strongest of the western Kansas forts. Only scattered bands of marauding Indians dared to come into the area and they fled as soon as they spotted white settlers. On April 1, 1867, two thousand men of the 7th Cavalry commanded by Colonel George Custer went into camp on the Smoky Bottom just west of the post. Wild Bill Hickok was attached as a scout, and a number of Delaware Indians accompanied the command as scouts, hunters, and interpreters.

In 1867 a cholera epidemic swept Fort Harker and other frontier military posts. The epidemic took the life of the wife of Brevet Major George Sternberg. Sternberg later became the Surgeon General of the United States and an authority on epidemic diseases. When news of the epidemic reached Fort Leavenworth, Sisters of Mercy, with two priests, were sent. One of the priests, Father Martyn, who administered to the sick was found dead midway between Fort Harker and the nearby city of Ellsworth. He had been stricken while on his way back to Harker.

When it was proposed in 1871 to abandon Fort Harker, the Kansas legislature passed a joint resolution of protest stating that Fort Harker was essential to the defense of the north-central Kansas frontier and, second, that it would be a great financial loss, since the buildings cost the government one million dollars. The fort was finally abandoned in 1873.

The guardhouse, which now houses Fort Harker Museum, is pretty much the same today as it was in the 1860s. It is a substantial two-story building at the west end of the parade ground. The upper floor is divided into two large rooms and six small cells. This floor, which is reached by an outside staircase, has three windows. The lower floor is divided into three rooms: a guardroom, a noncommissioned officers' room, and a small tool room. At present the cells are gone from the upstairs, but it is planned to restore them eventually. The bars on the upstairs windows are original, those downstairs are copies. A bar missing from one upstairs window is said to have been the means of escape by Chief Santanta who was a prisoner here. The guardhouse sits at a thirteen-degree angle from the streets near it. This is because it was laid out by magnetic compass. The fort was named by General Hancock for Captain Charles G. Harker, killed at the battle of Kennesaw Mountain during the Civil War. Most of the Civil War generals were stationed at or visited Fort Harker. General Sheridan was in command of Harker when Forcyth and fifty civilian scouts set out from there for the Indian territory and were massacred at the Arickaree fork of the Republican River.

The museum exhibits uniforms of men from this area who served in recent wars. A horse-drawn ambulance is exhibited. It is the same type as was used to bring wounded men in from the Indian wars and also served as transportation for families being moved from one post to another.

Fort Harker is located on West Ohio Street in Kanopolis, Kansas, one and three-quarter miles south of K-140 Highway in the center of Kansas. It is open all year from 10:00 A.M. to noon and from 1:00 to 5:00 P.M. It is closed on Sunday morning. No admission fee.

Fort Hays

HAYS, KANSAS

SHORTLY AFTER the Civil War railroads and settlers began pushing into central and western Kansas with increasing intensity. To provide protection against hostile Indians, who attacked travelers, railroad builders, and settlers alike, the federal government in 1865 established three new military posts in Kansas: Fort Dodge on the Santa Fe Trail, Fort Wallace and Fort Hays established on the Smokey Hill trail, a stagecoach road to Denver and later the route of the Union Pacific Railroad, Eastern Division.

The fort was first named Camp Fletcher and manned by Company G, Second Cavalry, and Companies F and G, First Volunteer Infantry, under the command of Lieutenant Colonel William Tamblyn. The camp was located fourteen miles southeast of present Hays. On November 17, 1865, the name was changed to Fort Hays in memory of General Alexander Hays, who had been killed in 1864 at the Battle of the Wilderness.

On June 7, 1867, a torrential rain set off flash floods that virtually destroyed the camp, and measures were at once taken to locate a more favorable site in the vicinity. The relocated fort, now just south of the city of Hays, bore little resemblance to its destroyed predecessor or to the typical military posts of the earlier eastern frontiers. There was no stockade or fortification wall. Instead, officers' quarters, barracks, headquarters, storehouses, and other buildings grouped around a parade ground constituted the outline of the new Fort Hays. A stone blockhouse was equipped with rifle slits, but the fort was never attacked and the building was actually used as a post headquarters and adjutant's quarters. The blockhouse was hexagonal in shape with two wings extending north and south. The guardhouse was also built of stone and was twenty-four feet wide and one hundred feet long, with a porch extending north and south. It furnished quarters for the Officer of the Guard, a guardroom, and a prison room with three cells. A stockade, at one time used for Indian prisoners, was attached to the guardhouse but was torn down in the 1870s. Except for a small bakery, all other buildings—officers' quarters, married enlisted men's quarters, barracks, hospital, storehouse, and other housekeeping buildings—were of frame construction.

Fort Hays was never a large post. Garrison strength ranged from a maximum of 567 men in 1867 to 12 in February of 1879. The number varied between these extremes depending upon Indian activity, averaging normally three companies, or about 210 men. During much of the post's active life both cavalry and infantry were stationed there. In the last few years it was primarily an infantry post. Elements of eight different cavalry and eleven infantry regiments, as well as artillery batteries, were based at the post at various times. Of the nineteen regiments, two were black.

Only a few weeks after the new Fort Hays was established the garrison was ravaged by cholera. The epidemic raged for two months, and when the dead from the post cemetery were disinterred and transferred to Fort Leavenworth in 1905 they included the bodies of 174 persons who died of cholera, most of them during the 1867 outbreak.

Indian attacks increased in the summer of 1867 throughout the vicinity and more cavalrymen were recruited and ordered to Fort Hays. Outside the fort the growth of Hays City coincided with the arrival of the railroad. It became

a rather wild frontier town with about 150 houses and 400 inhabitants, most of them single men and women who were not overly moral. It was filled with saloons and dance halls. Wild Bill Hickok was sherriff in 1869 and served only five months. He left town hurriedly after a brawl with some 7th Cavalry troopers, one of whom was killed.

Indian uprising erupted again in the summer of 1868 and continued until 1870. General Sheridan came west and made Hays his headquarters for several months. According to the War Department statistics the two years 1868-69 were among the worst in the history of Plains warfare. There were twelve engagements between Indians and the military. Thirty soldiers, 5 scouts, and 293 Indians were killed, and 37 soldiers, 14 scouts, and about 250 Indians wounded. Civilian casualties included 116 killed, 16 wounded, 32 scalped, and 7 captured. Four wagon trains, twelve stagecoaches, and twenty-one houses were attacked, burned, or damaged.

By the late 1870s, most of the region had been settled by the advancing white civilization. As the white population increased, the great buffalo herds became smaller and this forced the Plains Indians to follow their main food supply to other areas. As the Indian menace declined, so did the need for a military post at Hays. By the mid-1870s commanders were recommending that the post be abandoned, but no action was taken for nearly fifteen years. Finally on November 8, 1889, the last garrison moved out.

In 1900 the federal government gave the site to the state of Kansas. In 1931 the site became Fort Hays Kansas State College, but a public park officially called Frontier Historical Park was plotted to include the remaining historic buildings at Hays. These numbered thirty-eight at the time the post was abandoned in 1889. Today only the stone blockhouse and guardhouse and one frame officers' quarters survive.

Fort Hays is on U.S. 183 Alt, four miles south of Interstate 70, in Frontier Historical Park at the south edge of Hays, Kansas. The fort may be visited during the summer months from 9:00 A.M. to 9:00 P.M.; on Sunday from 1:00 to 9:00 P.M.; during the winter months from 10:00 A.M. to 6:00 P.M.; on Sunday from 1:00 to 6:00 P.M. No admission fee.

Fort Laramie

FORT LARAMIE, WYOMING

OUTSTANDING among the army posts that influenced the conquest and settlement of the West, Fort Laramie was a center for trade, diplomacy, and warfare on the northern plains frontier from 1834 to 1890. First as a private post, then later as a post of the American Fur Company, the fort engaged in the fur trade of the Rocky Mountain area. After 1849 the military fort here protected the westward migration to Oregon, Utah, and California. Until the fort was abandoned in 1890, its command was involved in the numerous campaigns and treaties designed to pacify the northern Plains Indians.

Trappers, Plains Indians, mountain men, covered-wagon trains to Oregon, Mormon pilgrims to Utah, "Forty-niners" to California, Pony Express riders, overland stage passengers, cowboys, and homesteaders—these were the colorful characteristics played out in all their varied roles in the old Fort Laramie. Fans of the Western movie have seen its replicas

The most important military outpost in the Wyoming of the West, Fort Laramie stood as the last major outpost for pioneers and emigrants on the Great Migration West. (Photo: Wyoming Travel Commission)

hundreds of times in hundreds of films. Fort Laramie has starred in most of them.

Not long after the Louisiana Purchase, Americans began to utilize the lucrative fur resources of the Rocky Mountains and the Great Plains. Late in 1812 Robert Stuart, carrying dispatches for John Jacob Astor eastward from Fort Astoria at the mouth of the Columbia River, reached the area of Fort Laramie and gave the first written description of the region. Ten years later William Ashley and Andrew Henry of St. Louis formed the Rocky Mountain Fur Company. This company trapped the country around the sources of the Platte, Green, Yellowstone, and Snake rivers. In 1830 Astor's powerful American Fur Company began to operate in the central Rockies seeking a monopoly. Four years later crushing competition forced the dissolution of Rocky Mountain Fur Company.

That same year, 1834, saw the establishment of Fort Laramie by William L. Sublette and Robert Campbell, two partners with more than a decade of experience as trappers, traders, and businessmen. Named Fort William in honor of Sublette, the post was shortly sold to a fur partnership which included Jim Bridger and Tom Fitzpatrick, two of the most famous "mountain men."

Bought in 1836 by the American Fur Company, Fort William was rebuilt of adobe in 1841 when its logs began to decay, and was renamed Fort John, but in common usage the post was known as Laramie, the name given to the river and the surrounding territory in memory of Jacques La Ramee, a trapper reported to have been killed about 1821 by the Indians on the banks of the stream known as Laramie River.

During its heyday as a fur-trading post, Fort Laramie was one of the major depots of the vast fur country of the Rockies. American traders and trappers, Indians, Frenchmen, and Canadians all gathered here, often en route to a summer rendezvous in the upper Green Valley of Wyoming. Bands of Sioux, Cheyenne, and Arapaho seeking to trade their furs camped near the post in large numbers. The homesteaders who passed in their covered wagons over this part of the Oregon Trail found the adobe-walled fort a welcome sight. There, they repaired their wagons and gave their animals a much-needed rest before the gradual ascent to the South Pass crossings of the Continental Divide.

In 1843 over 1,000 people passed Fort Laramie. For the next two decades more than 200,000 emigrants stopped there on their ways to Oregon, Utah, and California. The Mormons, in 1847, when they were driven out of Missouri and Illinois by persecution, rested at Fort Laramie on their way to Salt Lake Valley. During the great 1849 gold rush, Fort Laramie was the most important way station on the Platte Route with an estimated 25,000 gold seekers passing through that year. The tide of emigration reached its peak in 1850 when 55,000 emigrants passed through Fort Laramie. When mail service was established between Independence, Missouri, and Salt Lake City during most of the 1850s, Fort Laramie was the division point: the mail stages from each end of the line met at the post on the fifteenth of the month. However, with the completion of the Union Pacific Railroad in 1869, the death knell was sounded for migration up the North Platte Valley. For a short period in the late 1870s, Fort Laramie was a station on the Cheyenne-Deadwood Stage Coach Line.

In the spring of 1849 Fort Laramie was bought by the U.S. Government from the American Fur Company and new buildings were erected. A great assembly of Indian tribes met with United States commissioners at Horse Creek near Fort Laramie in the summer of 1851, and a treaty was signed whereby the

Indians agreed not to molest wagon trains and to permit the stationing of troops along the trail, in return for which the government was to pay them $50,000 annually in goods and respect their rights to traditional hunting grounds. This peace lasted until 1854 when the commander of Fort Laramie sent Lieutenant John L. Grattan with thirty men to a Sioux Camp to apprehend a cattle thief. The Indians refused to surrender the offender and in the dispute that followed the Indians annihilated Grattan's detachment. This is conceded to be the beginning of the war between the Plains Indians and the U.S. Army, which lasted without any interruption for thirty-five years.

In June 1866 peace commissioners of the federal government met with some 2,000 Sioux and Cheyenne at Fort Laramie to obtain their consent to build the Bozeman Trail to Virginia City, Montana. In the midst of the conference, troops under Colonel Henry Carrington appeared and Red Cloud and his tribesmen withdrew resentfully. Carrington's men constructed three forts along the Bozeman Trail, but Red Cloud's warriors kept them in a state of siege. A treaty with the Indians in April 1868 resulted in the abandonment of the Bozeman Trail forts, and this treaty marked the end of Fort Laramie as the great trade center of the Sioux. They were forbidden to come to the post. The last military activities of the post were in 1875–76 when the gold rush to the Black Hills brought hundreds of prospectors to the fort. The last great stand of the Sioux against the white men was against the prospectors. General George Crook's operations from the fort broke the resistance of the Indians.

With the Indians subdued, the fort was no longer needed to serve an essential military function. In 1890 the Army ordered the fort's abandonment and the following spring the last troops departed. When the troops marched away the sixty-five fort buildings were auctioned off to the homesteaders. Many were dismantled. Several were preserved as dwellings, barns, etc. Many years passed, however, before the historic importance of the old fort was recognized. Serious efforts to save old Fort Laramie were launched in 1930. In 1937 Wyoming purchased the site and donated it to the federal government. It was proclaimed a National Monument in 1938 and redesignated a National Historic Site in 1960.

Eleven original structures are restored and may be seen while the standing ruins of ten others and the outlines of many more are readily discernible. The oldest of these historic buildings—Old Bedlam, the magazine, the trading post section of the sutler's store, and the ruins of the original guardhouse—date from 1849–52. Among the other notable structures preserved or restored are the cavalry barracks, commissary warehouse, three officers' quarters of the 1880–84 period, and the middle and late period guardhouses. Conspicuous ruins are those of the hospital, administrative building, the noncommissioned officers' quarters, and several officers' quarters.

Fort Laramie National Historic Site is three miles southwest of the town of Fort Laramie, Wyoming, which is on U.S. 26 between its junctions with U.S. 85 and 87. It is about twenty-three miles west of Torrington, thirteen miles west of Lingle, and thirteen miles east of Guernsey, Wyoming. The area is open throughout the year. There are no facilities for camping. No admission fee.

Fort Laramie (Photo: Wyoming Travel Commission)

Fort Larned

LARNED, KANSAS

DURING THE EARLY nineteenth century, a thin strip of dusty trail on an unbroken stretch of prairie lay between the frontier settlements of western Missouri and the adobe town of Santa Fe, New Mexico. This eight-hundred-mile-long strip was called the Santa Fe Trail. Along this long and hazardous road, commerce and settlers traveled, fought, and died—it became a great artery of commerce that held limitless profits for those who survived its perils. It was along this trail in Kansas that Indians, tradesmen, settlers, and soldiers fought and died. Men like Kit Carson and Buffalo Bill Cody earned their legendlike reputations along its length, and military outposts were thrown up to escort and protect the wary traveler.

It was along this trail that Fort Larned became famous as well as Pawnee Rock, a promontory rising up off the floor of the plains about fifteen miles to the east. New Franklin, Missouri, was the original eastern terminal of the Santa Fe Trail, then successively Arrow Rock, Independence, and finally Westport (now Kansas City). Three eras unfolded in the development of the trail: The first was symbolized by the pack mule, the second by the ox team, and the third by the horse and mule.

Enlisted men's barracks, Fort Larned

The huge wagons for transporting goods, government supplies, and settlers were loaded with six to seven thousand pounds at Independence or Leavenworth. A train consisted of twenty-six wagons. A full crew was twenty-nine men; twenty-six drivers, boss, assistant boss, and an extra. The two bosses rode mules. To protect the cargoes and the crews of these impressive wagon trains the government established forts that constituted a chain of defense from the Missouri River to Santa Fe, New Mexico. The first fort after leaving Council Grove, Kansas, was Fort Zarah (two miles east of the Great Bend) on the Arkansas River. Next in order—to the southwest along the trail route—was Fort Larned, and beyond it Fort Dodge. Fort Larned was considered the most important outpost on the old trail.

Fort Larned was established by Major Henry W. Wessells, October 22, 1859, as "Camp on Pawnee Fork." During the next year the name was changed twice—first to Camp Alert (because of imminent Indian attacks) and then to Fort Larned, in honor of Colonel B. F. Larned, then paymaster of the United States Army.

Orders from Fort Leavenworth for the new fort were explicit: "Garrison the post . . . escort the mails, east and west . . . guard Santa Fe crossing at Pawnee Fork." During the ten years of the Indian wars that followed, Fort Larned's functions became far more ramified. Various businesses—a sutler's store, saloon, stageline station, fur-trading camp operated by Dave Butterfield of the express company, and a blacksmith shop—all bunched close to the fort quadrangle to serve both troops and travelers.

Scores of dugouts and half-soddies dotted the south bank of Pawnee Fork (now Pawnee River) where many employees lived and trail travelers bedded down in severe weather.

The fort was never palisaded but an earthen breastwork completely surrounded the build-ings in time of trouble. The fort itself was built of adobe with earthen roofs and then of sandstone when the present fort was constructed in the middle 1860s just southwest of the old post. Fort Larned is probably one of the best preserved old military posts in America today.

The fort was a scene of intense activity. By 1862 about $40 million worth of traffic moved along the Santa Fe Trail in a single year, and as many as twelve thousand teamsters and other travelers annually passed by the post. The scene around the fort was lively and colorful—teamsters swinging ponderous Conestoga wagons into line with noisy profanity helping their efforts, Indians gambling and sprinting their fast ponies in impromptu races around the post, and troopers going about their military duties. In 1860 the fort was made the agency of the Cheyenne, Arapaho, and Plains Apache Indians, and there were often as many as three thousand Indians camped nearby.

Indian turmoil began almost immediately after the fort was established. There were Indian attacks against nearby settlers. Wagon trains were ambushed and attacked, with the Indians openly defying the nearby fort and its small complement of soldiers. At the outbreak of the Civil War Fort Larned was western Kansas' lone military outpost. Built to quarter about two hundred soldiers—its average complement—it often had green volunteers and later even Confederate prisoner troops. The Confederate soldiers had been paroled from the East to fight the Indians. Usually they were so weakened from imprisonment they could march no more then eight miles a day. Yet the men often fought with distinction. During the bloody Indian War the War Department designated Fort Larned as a rendezvous point for all wagons proceeding westward. Wagon train after wagon train piled up at the post awaiting escorts. At one time, Fort Larned witnessed perhaps the most colorful scene in pioneer history—one thousand huge wagons, traveling in columns of four, moved in a vast line away from Fort Larned toward Santa Fe. No wagon train in all the West, before or after, ever attained such awesome proportions.

For more than six years Fort Larned had undergone every kind of Indian attack and had hung on. The Indians remained fighting and defiant. The struggle for the plains continued through 1866, and the next year the War Department decided to force peace by a display of force. A two-thousand man expedition under General Winfield Scott Hancock marched to Fort Larned in the mightiest array of manpower the plains had ever seen. Second in command was General George Armstrong Custer. The expedition included seven companies of infantry, a pontoon trail, a battery of artillery, and six companies of cavalry. They were accompanied by reporters and artists. Their news dispatches made Fort Larned familiar throughout the United States and Europe. After a number of ineffectual attempts at peace talks with the Indians, finally, in October 1867, the Medicine

Enlisted men's barracks, Fort Larned (Photo: Kansas State Historical Society, Topeka)

Lodge Peace Treaty was made at Medicine Creek Lodge, Kansas, about eighty miles southeast of Fort Larned.

Even during the Indian turmoil, Fort Larned found time for several social activities. Full-dress dinner parties were given officers on inspection trips. Numerous quiltings, taffy pullings, and cockfights took place. At the occasional dances, everyone joined in for the quadrille, polka, or scottische—to the music of a guitar or cornet.

As Indians were removed to reservations, the necessity for frontier posts slackened. Finally, all the troops at Fort Larned were transferred to Fort Dodge.

In 1882 Congress approved the bill to authorize the sale of the military reservations, and on March 13, 1884, Fort Larned was sold to the Pawnee Valley Stock Breeders Association and used as a stock ranch. On June 28, 1902, it was purchased by E. E. Frizell. During the succeeding years it became one of Western Kansas' most important cattle and grain ranches. E. D. Frizell, eldest son of the senator, guided the fort's ranching activities until his death in 1956.

Today, Fort Larned, under the ownership of Mrs. E. D. Frizell and her son Robert Frizell is leased to the Fort Larned Historical Society, a nonprofit organization. It was opened to the public May 19, 1957. It stands today as the symbol of a dramatic military past.

The visitor today can see the nine massive stone buildings that constitute Fort Larned. Built in the 1860s to replace adobe structures, they are still preserved in the original state,

around the garrison parade ground. They remind the visitor of the more than one hundred battles and skirmishes fought out of Fort Larned with Indians. They remind one that Buffalo Bill, Wild Bill Hickok, and Kit Carson were frequent visitors and that more than five hundred troopers, plainsmen, and Indians were killed or wounded within thirty miles of Fort Larned. This dramatic past is graphically illustrated in the museum, where Indian artifacts of the area and hundreds of military and pioneer items are displayed. Here, also, are the guns that won the West, including the "Old Colts," Sharps, and Springfields.

The original blacksmith and wheelwright shop is preserved with forge, bellows, and tools. Thousands of horses, mules, and oxen were shod here. One can see the Old Escape Tunnel used to obtain water from the river and the gunports used by troops to protect the fort on the one side where the Pawnee River offered no barrier. The officers' quarters are carefully refurnished with antique pieces and pioneer relics. An Old Harness Shop displays early saddles, and there is a museum of horse-drawn vehicles.

Fort Larned is located six miles west of Larned, Kansas, close to U.S. Highway 156. It is easily reached by a short State Highway, 242. It is only six miles from U.S. 56, the Old Santa Fe Trail route, and only four miles east of U.S. 183, a major north-south highway. It is open to the public from 7:00 A.M. to 8:00 P.M. in summer and 9:00 A.M. to 5:00 P.M. in winter. No admission fee.

Fort Leavenworth

LEAVENWORTH, KANSAS

Founded in 1827 to protect caravans on the Santa Fe Trail and to help maintain a "permanent Indian frontier," Fort Leavenworth was the central fort in a chain ranging from Fort Snelling in Minnesota to Fort Jesup in Louisiana that the Army established in the 1820s and 1830s. From that time to the present it has been a major U.S. Army installation. A significant post in the trans-Mississippi West, it figured prominently in the Indian Wars of the Great Plains, the Mexican War, and the Civil War, and later became a major training center.

As early as 1824 citizens of Missouri petitioned Congress for the activation of a military post at the Arkansas Crossing for the protection of the traders of the Santa Fe Trail. Three years later the Secretary of War ordered the erection of a fort near the western boundary of Missouri to meet the petitioners' needs and quell Indian disturbances. Colonel Henry Leavenworth selected a site to the north of Arkansas Crossing, and his troops built a post called Cantonment Leavenworth, later Fort Leavenworth. Strategically located on the Missouri River near the eastern terminus of the Oregon and Santa Fe trails, for many years it was a key frontier post. From 1827 to 1839 it was headquarters of the Upper Missouri Indian Agency, which had jurisdiction over all the tribes in the Upper Missouri and Northern Plains region, and was the scene of many conferences and treaty councils. Exploring expeditions that used it as a base of operations between 1829 and 1845 included Major Bennett Riley's expedition along the Santa Fe Trail, Colonel Henry Dodge's expedition to the Rocky Mountains, and Colonel Stephen W.

Kearny's expedition to Cherokee country and to South Pass and the Rockies.

During the Mexican War the fort was the departure point and supply base for General Kearny's "Army of the West," which occupied New Mexico and California. Following the war the fort was the chief supply depot for army posts in the West, and in 1854 served as temporary capital of the Territory of Kansas. During the Civil War the Confederates twice threatened the fort, which after the war continued to be a major supply depot. From 1860 to 1874 it was an ordinance arsenal; from 1874 to 1878 the quartermaster depot for the Military Division of the Missouri; and after 1881 a school for infantry and cavalry, reorganized in 1901 as the General Service and Staff School. In the twentieth century it has served as an officer school, induction and training center. It is a Registered National Historic Landmark relating primarily to Indian-military affairs in the trans-Mississippi West.

Among the noteworthy historic structures of the fort are the post chapel, built in 1878, the original home of the General School for Officers; the enlisted men's barracks, constructed between 1881 and 1889; the Syracuse House, built in the late 1860s; and a portion of a wall of the original fort. One of the old cavalry stables now serves as a transportation museum.

Fort Leavenworth is located in Leavenworth County on the eastern edge of Leavenworth, at Reynolds and Gibbon Avenues. It is open from Monday to Saturday, 10:00 A.M. to 4:00 P.M.; Sunday and holidays from noon to 4:00 P.M. No admission fee.

U.S. Army Command and General Staff College, Fort Leavenworth (Photo: U.S. Army Photograph)

Fort Mackinac

MACKINAC ISLAND, MICHIGAN

In 1780, during the American Revolution, the British Lieutenant Governor Patrick Sinclair began to have misgivings about the vulnerability of Fort Michilimacknac. He decided that it could not be defended against American attacks and planned to take advantage of the natural water barriers surrounding Mackinac Island and build a new fort there. He dismantled several of the fort's structures and moved them to a new site on the island—a high limestone bluff overlooking the harbor. Here he built Fort Mackinac. In addition to the newly transferred structures he had built an officers' quarters out of the native limestone and placed within the stockade and ramparts three blockhouses, a hospital, wooden officers' quarters, a guardhouse, commissary, post headquarters, quartermaster's storehouse, enlisted soldiers' barracks, bathhouse, and commanding officer's quarters. Adjoining the fort was the surgeon's quarters and many years later, two cottages.

The new fort was responsible for the island's development as the great fur-trading center of the northern Great Lakes. Ever since its first day, life on the islands has revolved around the fort. After the Revolutionary War the Americans won the fort by treaty, but in the early days of the War of 1812 the British recaptured it from a small army garrison. When the status quo was restored after 1815, the United States troops again took possession, this time to stay until 1895.

Even before the last soldier left the fort, Mackinac Island, with its lovely scenery and excellent climate, was attracting more and more visitors. To ensure that the natural attractions were to be enjoyed for future generations, the government established a National Park on Mackinac Island. In 1895, when the fort's garrison was withdrawn, the United States turned over the park and the fort to the State of Michigan. The Mackinac Island State Park Commission was specifically charged with the preservation of the old fort. The oldest existing building and most impressive structure in the fort was Governor Sinclair's stone quarters. The other buildings in the fort gradually deteriorated, and by the 1950s most of the buildings were badly in need of repair and some were in hazardous condition. The opening of the new Mackinac Bridge in 1957 greatly increased the flow of visitors to the island, and a massive movement was begun to restore the fort and its surrounding grounds.

Architects, museum designers, research historians, curators, archaeologists, artists, and diorama makers were employed. Artifacts and excavations of sites provided invaluable data for the restorers. Much information on the buildings was gained through a study of documents and pictures. Parts of the buildings were repaired; other structures were restored as close to their original appearance as possible. Of the fourteen buildings in the fort, the north, east, and west blockhouses, soldiers' barracks, quartermaster's storehouse, post headquarters, schoolhouse, hospital, commissary, post commandant's house, guardhouse, and officers' wooden quarters are completed. The restoration of the officers' stone quarters is in progress. Historical restoration on the island covered other projects outside the fort. The Indian dormitory built in 1836 to house Indians who came to the island to confer with the Indian agent and the restoration of the Benjamin blacksmith shop are two of the projects.

The stone ramparts, the south sally port, and the officers' stone quarters are all part of the

North blockhouse, Fort Mackinac

Highland view, Fort Mackinac

original fort built nearly two hundred years ago. The other buildings are of more recent origin, dating from the late 1790s to 1885. Mackinac Island is in Lake Huron, three miles east of the southeastern tip of the upper peninsula of Michigan and is accessible by ferry from St. Ignace and Mackinaw City. The fort is open every day from mid-May to October 1 from 9:00 A.M. to 6:00 P.M. Adults $2.00. Children under twelve free.

Fort Michilimackinac

MACKINAW CITY, MICHIGAN

BUILT BY THE FRENCH in 1715 Fort Michilimackinac was the most important fur trade center in the West. Hundreds of French traders, *coureurs de bois* and *voyageurs*, as well as British and Indians traded here until 1780, when the fort was abandoned. It was here in 1755 that an Indian war party assembled, led by Charles Langlade, and played a key role in the rout of British General Braddock's expedition in far-off Pennsylvania. In 1761 British troops occupied the fort after France surrendered her North American empire to the English.

The tide was turned for the British in 1763 during Pontiac's Indian uprising when most of the British garrison was killed and taken prisoner. Ezekial Solomon, Michigan's first Jewish settler, and other traders saved themselves by hiding in garrets of French homes. In the following year the British reoccupied the fort.

In 1766 Major Robert Rogers of Rogers' Rangers fame was appointed commandant. He sent out an expedition to seek a northwest passage to the Pacific. Unfortunately the Northwest Passage was more mythology than real, and this nonexistent passage to the Orient was the cause of his being arrested on charges of treason in 1767. He was thrown into the guardhouse and then taken to Montreal for trial. He was acquitted but never fully recovered from the scandal.

During the Revolutionary War, Indian parties were sent out from the fort by the British to raid western outposts of the Americans and their Spanish allies. In 1779, fearing that the fort could not be defended against possible American raids, Lieutenant Governor Patrick Sinclair built a new fort on Mackinac Island and abandoned Fort Michilimackinac.

In 1904 the site of the old fort was acquired by the Mackinac Island State Park Commission, but it wasn't until 1959 that a restoration, or, more correctly, a reconstruction, of the fort began. Although the fort existed sixty-five years and at its height during the British occupation protected a thriving community of fur traders, artisans, soldiers, and their families, what had not been transported to Mackinac Island by the British in 1780 had disappeared through fire and rot. Once the fort had over thirty well-built log houses inside a twenty-foot-high stockade as well as over one hundred houses outside its walls, but within a few years all that remained of that fort civilization lay buried under a cover of forest humus and shore sand.

In 1960 the Mackinac Island State Park Commission undertook to reconstruct the eighteenth-century fort and create a historical environment in the surrounding park. All extant records were studied as well as the site itself. Contemporary plans showed lines of wall pickets, land and water gates, bastions, a wharf, and over thirty houses and their gardens. Pa-

French traders' houses at Fort Michilmackinac

Fort Michilmackinac and view of Mackinac Bridge

pers in the British Museum furnished descriptions of the fort buildings. Archaeologists were also digging up the site and uncovered foundations, fireplaces, remnants of log walls, wells, and basements, which revealed dozens of details of the original fort. The actual reconstruction was done by skilled craftsmen using a combination of old and new building techniques. Stockade walls were erected; four corner bastions and four blockhouses were built. The soldiers' barracks were rebuilt, as was the King's storehouse, a British trader's home, the Church of Sainte Anne de Michilimackinac, the priest's house, commanding officer's house, a blacksmith shop, and a complex of French traders' dwellings. The combined research of archaeologists and historians and the labor of skilled craftsmen have resulted in one of America's most completely authentic reconstructions.

Fort Michilimackinac in Mackinaw City can be reached by highways U.S. 23, U.S. 31, and the lower Michigan routes I-75. It is open from May 15 to October 15, from 9:00 A.M. to 5:00 P.M. These hours are subject to change early and late in the season. Adults $2.00. Children under twelve free.

96

Old Fort Nisqually

TACOMA, WASHINGTON

FORT NISQUALLY was one of the Hudson's Bay Company's furtrading forts that was first constructed in 1833 along the bluff about two miles north of the Nisqually River. In 1843 it was rebuilt one mile farther inland adjacent to the Sequalitchew Creek.

In its most active period brigades of fur-laden pack horses from the rich newly tapped

Fort Nisqually is a replica of an early day Hudson's Bay Company post (Photo: Washington State Travel Photo)

Cascade territory between the Fraser and Columbia rivers deposited their pelts at the fort in exchange for goods and money. An honored place in the fort's history is given to Dr. William F. Tolmie of the Hudson's Bay Company. It was he who received and fed the weary settlers, exhausted from their long overland trek to Fort Nisqually and the lands they sought to settle. He extended credit to them for the purchase of supplies and equipment for their primitive farms hewn from the surrounding wilderness. In payment he allowed them a fair and generous price for their produce. In a way the restored fort is a memorial to the good doctor.

In 1869 the Hudson's Bay Company sold their holdings in the Pacific Northwest and the fort was preempted by the last factor, Edward Huggins, who lived in the fort until he died in 1906. The DuPont Company acquired the holdings in 1906, and when they were about to destroy the remnants of the old fort, a move was made to preserve at least the granary building which was determined to be the oldest extant building in the state of Washington.

In 1933 the Tacoma Young Men's Business Club, with the help of the Metropolitan Park Board of Tacoma, obtained the buildings of the fort and determined to restore the old fort. They relocated the restored fort in Point De-

fiance Park in northwest Tacoma. The log palisades, bastions, and several buildings have been authentically restored. Ninety percent of the granary had original material when restored. Thirty percent of the factor's house and the siding of the store building was salvaged from various sources about the old fort. They equipped the old blacksmith shop with a forge, leather bellows, anvil and tools, and other equipment as it was when the original wrought-iron hardware was hand-forged. Some of the original wrought-iron hardware has been incorporated in the other restored buildings. The bakery, with its dough-raising troughs, brick ovens, and bread-turning paddles is being restored.

Old Fort Nisqually's location in Point Defiance Park overlooks Puget Sound's Nisqually Reach, the Olympic Mountains, and the new Tacoma Narrows Bridge. The park is in northwest Tacoma. The roads in the park are clearly marked. During the summer, the fort is open every day from 8:00 A.M. to 8:00 P.M. and the small museum in the factor's house is open from noon to 6:00 P.M. During the fall and winter the fort is closed all day Monday and the museum hours are shortened to 1:00 to 4:00 P.M. No admission fees.

Fort Recovery

MERCER COUNTY, OHIO

THE MOST COMPLETE ROUT of a U. S. Army was suffered on the grounds of Fort Recovery before it was ever built. It was November of 1791 and a light snow covered the ground of General Arthur St. Clair's encampment on the Wabash some thirty miles north of Fort Jefferson. It was half an hour before sunrise and the troops had just been dismissed from

parade. Out of the stillness of the surrounding country the Indian forces of Little Turtle and Blue Jacket attacked from all sides, and before the soldiers could load their muskets they had been defeated. Before the day had ended three quarters of the troops were either dead or wounded and the survivors made their way to Fort Jefferson. During the winter the army

Statue of frontier soldier, Fort Recovery Monument (Photo: Ohio Historical Society)

One of two restored blockhouses, Fort Recovery (Photo: Ohio Historical Society)

returned to bury the bones of the massacred but would not attempt an occupation of the field.

Two years later, Major General Anthony Wayne, a hero of the American Revolution—Mad Anthony to his contemporaries—determined to reoccupy and hold the scene of St. Clair's defeat. On December 23, 1793, he detached a force of men under Major Henry Burbeck to the battleground with orders to construct a regular four-blockhoused post. Each of the blockhouses was to be twenty feet square and they were to be connected by pickets. He called the new post "Recovery." The fort was built with one-story blockhouses in four or five days, and on December 28 Wayne thanked Burbeck and his officers and troops for "repossessing the field of battle and erecting thereon Fort Recovery—a work impervious to savage attack." In March 1794 the fort commandant, Captain Alexander Gibson, added a second story to each of the blockhouses.

Just before summer of that same year the Indians met at Nuquijake Town near Au Glaize to plan another campaign to throw back the white invaders. Their strategy consisted of an attack upon Fort Hamilton, main supply depot for the advancing American army, to be followed by assaults on the forward posts—St. Clair, Jefferson, Greene Ville, and Recovery—one by one.

The army set out southward. When it had arrived just east of Fort Recovery, a change of plans turned the Indians on this post. News had come of a supply convoy making its way from Green Ville to Recovery.

Following the same pattern they used successfully against St. Clair, the Indians attacked the supply detachment encamped just outside the walls of Recovery, early on the morning of June 30, 1794. This time a complete encirclement was impossible because the fort itself cut off effective concerted operation.

The battle was a long and bloody one. The Indians had the advantage for only the first brief moment of the initial attack when they killed Major William McMahon, chief of the detachment troops, and several of his officers and men. Falling back yard by yard, the major portion of the detachment withdrew into Recovery's walls. From time to time, as the smoke cleared, the Americans could distinguish the red coats of the Indian's British allies in the background encouraging the attack. As the day drew to a close with all the Americans safely inside the stockade, the firing was limited to sporadic shots, and as night fell, the Indians picked up their dead and wounded. The next morning the attack was renewed, but soon the Indians gave up the fight and left.

Here, the importance of a small fort was made dramatically clear. Where one army was butchered, a second force withstood what was perhaps the largest Indian army ever to gather against American troops in the Indian Wars.

Today, in the center of the small Mercer County town named Recovery, a tall stone obelisk marks the final resting place of the dead of St. Clair's defeat and of Fort Recovery's victory. A few short steps to the west are two large blockhouses and a connecting stockade designating the site of this famous fort. Fort Recovery can be reached on Road 119 driving west of St. Henry or Road 49 driving north from Lightsville. It is open during the daylight hours. Adults 50¢.

Fort Ridgley

FAIRFAX, MINNESOTA

IN THE SPRING of 1853 the steamboat *West Newton* journeyed from Fort Snelling up the Minnesota River farther than any boat had previously ventured in those waters. Carrying soldiers and their families, carpenters, and supplies, the 150-foot-long vessel was bound for a high plateau above the river in the northwest corner of what is now Nicollet County. The men had been assigned to build a fort at the edge of the Dakota reservation, which had been established by treaties with these Indians in 1851.

The site was surrounded by deep, wooded ravines, and some of the three companies of soldiers who had been transferred from Fort Snelling and Fort Dodge to help build the fort thought "it was the worst place they ever beheld." The fort was called Ridgley in honor of three men of that name who had died during the Mexican War.

By 1855 Fort Ridgley was complete, although the original plans for the post had been altered. Only the commissary and barracks were built of stone, while the remaining structures were of wood; no stockade was built, and no well ever dug. Unprotected by a stockade and situated on an open prairie plateau, Fort Ridgley was ill prepared to withstand attack. Yet fewer than two hundred volunteer soldiers and civilian refugees successfully defended the fort against heavy odds in two battles.

In 1851 the government purchased 24 million acres of land and confined the Dakota (Sioux) to a narrow reservation along the Minnesota River. They wanted to change this tribe from a hunting to a farming society. A crop failure in 1861 brought many Dakota near starvation the following spring, and when the yearly payment for their ceded land, due in June, had not arrived by August, the Indians had reached the breaking point. On August 17, 1862, four young braves, acting on a dare, shot five white settlers near Acton in Meeker County. Fearing severe punishment, the Indians decided on war. News of the war arrived at Fort Ridgley on August 18, when a scout reported the Indians' attack on the Lower Sioux Agency. Captain John S. Marsh, commander of Fort Ridgley's troops, and some forty-five soldiers left for the agency. On the way they were ambushed at Redwood Ferry. Marsh drowned in an escape attempt and more than half of his men were killed. On that same day, the $71,000 gold annuity arrived two months behind schedule and one day too late to prevent a war.

The first battle at the fort did not come until August 20. There were about 180 men inside the fort, commanded by Lieutenant Timothy J. Sheehan, who had succeeded Captain Marsh. Little Crow, the Dakota leader, had gathered about four hundred warriors in the deep, wooded ravines around the fort. The fighting lasted about five hours, but the fort's fire kept the Indians at a safe distance. Rain fell on August 21. When the Indians returned on August 22, they had eight hundred warriors, but withdrew after eight hours of fighting. The five cannon in the fort frightened the Dakota and prevented them from ever gaining entry into the fort. The siege of the fort was lifted when Colonels Samuel McPhail and Henry H. Sibley led troops into the fort on August 27 and 28. In the two Fort Ridgley battles, three soldiers died and thirteen were wounded. Indian casualties were estimated from a few to over a hundred.

The defenders held the fort against a force twice its size because of their skill in handling the cannon. It broke up attacks before they really got started. The Indians retreated slowly up the Minnesota Valley, fighting several battles with soldiers and settlers. A victory for Sibley's forces at Wood Lake on September 23 ended the organized warfare by the Dakota in Minnesota. On September 26 the Indians freed 269 captives at Camp Release. Thirty-eight Indians charged with participating in the war were hanged at Mankato on December 26.

The role of Fort Ridgley and its garrison in the Dakota War was a decisive one. They settled the outcome of the war before it actually ended.

Fort Ridgley soon developed into a self-sufficient community, although from 1853 to 1861 there were never more than three hundred soldiers and civilians living there at one time. When the Civil War began in 1861, volunteer troops replaced regular army troops at the fort until the regulars returned in 1865 after the war. Fort Ridgley was abandoned in 1867 after the Dakota had been expelled from the state and the need for the military post no longer existed.

By the early 1880s hardly a trace was left of the original fort. Settlers dismantled the buildings and used the land as homestead until 1896 when the state began purchasing it for use as a state park. When archaeological excavations began at the site in 1936, only part of the stone commissary and a log magazine used as a shed on a nearby farm remained. The fort has been partially restored. Visitors can see the restored commissary, and the magazine has been repaired and returned to the park site, as well as the excavated and marked foundations of other buildings. The stone commissary houses exhibits explaining Fort Ridgley's history and the part it played in Minnesota's Indian War.

Fort Ridgley is located in Fort Ridgley State Park and is accesible from State Highway 4, seven miles south of Fairfax. The fee to enter the park is $1.50. It is open daily from 10:00 A.M. to 5:00 P.M., May 1 to October 15.

Fort Robinson

CRAWFORD, NEBRASKA

THE FIRST plainsmen were prehistoric Indian hunters of ten thousand years ago who established hunting as the pattern for their way of life. Their successors in Western Nebraska followed the same patterns thousands of years later when their life consisted of hunting buffalo and other game for food, shelter, and clothing.

The Sioux were inheritors of this tradition. They were the nineteenth-century occupants of the Fort Robinson area. Their nomadic existence from hunting area to hunting area made them skilled horsemen and fierce warriors. When white settlements threatened their hunting grounds, their resistance was well organized and fanatical.

In 1868 a treaty was signed which guaranteed the Sioux and other tribes food and supplies in exchange for lands ceded to the United States. The annuity goods granted to the Oglala Sioux by this treaty were issued at the Red Cloud Agency which was located on the Platte River. Sioux, Cheyenne, Arapaho, and Miniconjou were regularly supplied at Red Cloud. Among the Indians living at the agency a small fraction was friendly to the white man, while the rest wavered from friendliness to hostility.

There were numerous skirmishes with hostile war parties, notably Crazy Horse, of Custer Battle fame, and these marauding bands

Blacksmith and harness shops, Fort Robinson

Old Post Headquarters (now Nebraska State Historical Society Museum), Fort Robinson

were pursued by cavalry patrols from nearby Fort Laramie. Events in February of 1874 brought the ever-present tension of the area to a breaking point. The murder of the government agent, mule stealing, ambush of U.S. troops, and other incidents made it clear that the Red Cloud Agency had become a powder keg. Troops were requested to protect the agency. Eight infantry and four cavalry companies marched from Fort D. A. Russell, Wyoming Territory, to Fort Laramie. There, four more companies of cavalry were added. When this formidable expedition reached the agency, it was decided that they establish there a military camp. On March 29, 1874, the name of the Camp Red Cloud Agency was changed to Camp Robinson in honor of Lieutenant Levi H. Robinson, who had been killed at Little Cottonwood Creek the previous month.

Indians were not the only persons contributing to the troubles at Red Cloud Agency which occupied the attention of the soldiers from Camp Robinson. Many horse thieves such as Doc Middleton's gang stole Indian mounts in the area. It was a road agents' rendezvous with men like Black Doak, Fly Speck Billy, Lame Johnny, Paddy Simons, and others frequenting the agency between their attacks on the stage-coaches traveling the Sidney-Deadwood and Cheyenne-Deadwood trails.

The campaign against the Indians was entrusted to General George Crook, the Army's supreme Indian fighter. In the spring and summer of 1876 battles were fought with Crazy Horse, Sitting Bull, and Black Moon. The winter led to new victories for General Cook and defeats for the Sioux. The final total of Indians who surrendered at Camp Robinson

and Sheridan reached almost 4,500. Crazy Horse himself was killed in a struggle at the Fort Robinson guardhouse in September of 1877.

Camp Robinson was renamed Fort Robinson in January 1878. It remained an important post, and its garrison was called upon in several Indian emergencies after the death of Crazy Horse. In 1879 Fort Robinson was the scene of the final episode of the epic Cheyenne Outbreak when the chief, Dull Knife, and his people escaped from their barracks prison and fought the post garrison on the parade ground and in the nearby buttes.

With final peace and the Indians on the reservations, ranchers and homesteaders moved into the area. In their sod houses and tarpaper shacks, these pioneers lived under frontier conditions well into the twentieth century.

After service in World War II, when it was used as War Dog Training Center and prisoner-of-war camp, the post was turned over to the U.S. Department of Agriculture, which in cooperation with the University of Nebraska operates the Fort Robinson Beef Cattle Research Station. In June 1956 the Nebraska State Historical Society opened the Fort Robinson Museum. The museum and other historically significant parts of the post have been transferred to the Historical Society. The Museum housed in the old post headquarters interprets the story of man's occupation of the western Plains.

Fort Robinson is located four miles west of Crawford, Nebraska, on U. S. Highway 20. It is open from April 1 to November 14, Saturday from 8:00 A.M. to 5:00 P.M.; Sunday from 1:00 to 5:00 P.M. No admission fee.

Fort Ross

SONOMA COUNTY, COAST HIGHWAY, CALIFORNIA

THE RUSSIAN-AMERICAN COMPANY, which established Fort Ross in 1812, was a trade monopoly chartered by Paul I (1796–1801) in 1799. Organized to exploit the fur resources of Alaska, its activities resembled those of the Hudson's Bay Company of Canada. The company established a number of posts in Alaska but were forced to seek an outpost in California

Commander's House, the second oldest wood structure west of the Rockies, Fort Ross (Photo: State of California, Division of Beaches and Parks)

because of the inadequate food supply in Alaska during the early 1800s. The uncertain arrival of supply ships was a factor, and a growing scarcity of fur animals, particularly sea otter and fur seal, was another. California, although controlled by Spain, was believed to be an answer to both problems.

The first independent Russian exploration and sea otter hunting expedition to California set sail from Sitka in 1808 under Ivan Aleksandrovich Kuskov. On January 8, 1809, the *Kodiak* dropped anchor in Bodega Bay, California. Temporary shelters were set up and some 130 hunters went to work. The *Kodiak* returned to Sitka in August of 1809 with 2,350 sea otter pelts. Since the northern California coast was unoccupied, it afforded all the features of a successful Russian outpost. In 1811 Kuskov returned to Bodega Bay and decided that Bodega was unsuitable as a permanent headquarters. In June of 1812 a crew of ninety-five Russians and forty Aleuts began work on a stockaded fort of redwood at the site of the Pome Indian village of Mad-shui-mui on an elevated coastal plateau overlooking a small harbor thirty miles north of Bodega Bay and thirteen miles northwest of the mouth of the Slavianka, or Russian, River. The fort was dedicated at "Ross" on August 13, 1812. The "Colony Ross" eventually became known as "Fort Ross." The word "Ross" is believed to come from the Russian word "Rus." The Russians now occupied a permanent trade base at Fort Ross and a harbor at Bodega Bay from which the needs of Alaska could be supplied.

Between 1805 and 1836 thousands of sea otters and California fur seals were killed and their pelts stored at Fort Ross for later shipment to markets in China, Manchuria, Siberia, and Europe. The Spanish and Mexican governments in California attempted to control foreign sea otter hunting by granting seasonal hunting permits. But by 1825, as a result of the

ruthless slaughter by the Russians, Americans, and British, few sea otters could be found. Between 1850 and 1937 California sea otters were reported to be extinct, but in 1937 a small herd was sighted off the Monterey coast. There are now several hundred California sea otters.

In 1821 the Czar issued an ukase closing the Pacific Coast north of San Francisco to all but Russian ships. This attempt at Russian control, plus the presence of the Russians at Fort Ross, was responsible for that part of the Monroe Doctrine of 1823 which stated the New World was no longer open to aggression by force and that European countries could not extend their holdings in it.

With the disappearance of the sea otter and fur seals, the Russians increased their efforts at agriculture and manufacture in their California colony but without success. By the end of 1839 the Moscow officials of the Russian-American Company ordered the colonists to sell out and return to Alaska. Negotiations for the sale were carried on with both General Mariano G. Vallejo of Sonoma and Captain John A. Sutter of New Helvetia (Sacramento). Sutter's offer was accepted on December 12, 1841. He was to pay $30,000 in produce and gold for the movable property and other assets of the Russian colony. Between 1841 and 1844 Sutter's men took down a number of buildings and removed the arms, equipment, and livestock that the Russians had left.

After 1845 the fort became the center of a large ranch, and the remaining buildings were used in various ways. After the collapse of the Chapel of Fort Ross in the earthquake of 1906, the fort site was purchased by the California Historical Landmarks Committee of San Francisco and presented to the state in the same year.

Today, one can see the restored chapel, which had retained much of its original structural detail. It is the oldest Russian Orthodox

Restored chapel, oldest Russian Orthodox chapel still standing in the U.S. (Photo: State of California, Division of Beaches and Parks)

chapel still standing in the United States and has been furnished much as it was in the early days. Also on view are the manager's or commander's house, a log building measuring about fifty-three feet by thirty feet, which stands in the west corner of the fort. The log walls, puncheon, ceiling, and most of the floors and partitions are original Russian construction. This building is believed to be the second oldest wood structure remaining west of the Rockies. The eight-sided blockhouse was re-built in 1917 but replaced in 1957 by a more authentic reproduction. The seven-sided blockhouse was restored in 1956. All the stockade has been restored. It stands twelve feet high and over nine inches thick, made from hand-hewed redwood timbers in the Russian manner of the original.

Fort Ross is located on California Highway 1, some thirteen miles north along the coast from Jenner. It is open daily from 10:00 A.M. to 5:00 P.M. The fee is $1.50 per car.

Fort St. Charles

LAKE OF THE WOODS, MINNESOTA

SURROUNDED ON THREE SIDES by Canada, at the tip of a northward projection called the Northwest Angle, is Magrussen Island and Lake of the Woods where Fort St. Charles, a fortress and fur-trading post, was erected in 1732 by a band of Canadian-French voyageurs led by Pierre La Verendrye. From this most western outpost of white men in America, subsidiary posts were established and a vast section of mid-continent North America was explored by La Verendrye and his party, which included three of his sons.

It was a small log fort, nestled in a lakeside clearing. It was named for the patron saint of the then-governor of New France, the Marquis de Beauharnais. It consisted of a double row of cedar posts forming an enclosure one hundred by sixty feet. Within the enclosure were a chapel, houses for the missionary and commandant, quarters for the voyageurs, a warehouse, and a powder magazine. There were two gates opposite each other protected by bastions.

Indians came to Fort St. Charles to barter their furs, and La Verendrye constantly urged them to keep the peace, particularly with the Sioux, who visited the lake from the south and raided the northern tribes. He induced some of the Indians to settle near the fort and taught them to sow corn and peas.

In May of 1736 his oldest son, Jean-Baptiste La Verendrye, succeeded him as commandant of the fort. In June of the same year, the food supply was dangerously low, consisting of only spoiled fish. The new commandant together with Father Jean-Pierre Aulneau decided to trek to Fort Michilimackinac to replenish their supplies. Accompanied by nineteen voyageurs in three canoes, they set out but never reached their destination. On an island twenty miles from the fort they were all massacred and beheaded by the Sioux. Later, the bodies of Jean-Baptiste La Verendrye and Father Aulneau and the heads of the voyageurs were returned to Fort St. Charles and buried.

In spite of the tragic death of his son and his party, Pierre La Lerendrye continued to do his utmost to keep the peace among the Indians and also continued his explorations. In 1749 he died in Montreal, and following his death, Fort St. Charles was abandoned. A few years later the French and Indian War and the surrender of Canada to the British ended French fur trade

*Fort St. Charles (Photo: Department of Economic
Development, State of Minnesota)*

in the west. With its meager official records entombed in government archives, Fort St. Charles was forgotten. Nature and the elements of the northland obliterated the stockade and buildings of the fort, and with the passing of 150 years, knowledge of its location remained only in Indian lore.

In 1889 three Jesuit priests gave a mission in Vendée, France. As it ended an old man named Aulneau approached the priests saying that they were the first Jesuits he had seen although generations before a member of his family became a Jesuit priest and was slain by Indians in North America. He showed them letters from this young priest preserved as heirlooms in his family for over 150 years. The missionaries had them copied and translated into English. Later the story and letters were published in the *Canadian Messenger*, a widely circulated Jesuit magazine. The Jesuits of St. Boniface College in Manitoba were greatly interested and in the summer of 1890 organized an expedition to search Lake of the Woods, locate Fort St. Charles, and recover the body of Father Jean-Pierre Aulneau. They were unsuccessful, as were other expeditions. Then, in 1908, another expedition sponsored by St. Boniface College found flat chimney stones and fireplaces. When they cleared trees and brush, they found the stumps of an old stockade. Fort St. Charles was found at last!

The site was trenched, mapped, and photographed. The skeletal remains of Jean-Baptiste La Verendrye and Father Jean-Pierre Aulneau and the nineteen skulls of the *voyageurs* were found and positive identification made. The bones were moved of St. Boniface College, and a cross with a wooden plaque was left to mark the site of the fort. During the following forty years the site was rarely visited and again reverted to the wilderness.

In 1951 the Minnesota 4th Degree Knights of Columbus decided to make the marking of the site and restoration of the fort a project. Work groups of members with only hand tools felled trees and cleared brush. They erected a Memorial Altar and in 1953 a chapel to shelter the Memorial Altar. In 1960 the stockade and bastions of the fort were rebuilt of cedar logs on the exact outline of the original fort and according to the descriptions found in old records. Within the stockade the known buildings with fireplaces have been indicated by foundation timbers, and the former graves of Father Aulneau and of his companions marked with proper headstones.

Fort St. Charles can be reached by boat from Warroad, the only American port on Lake of the Woods. There is also regular boat service from Kenora to Flag and Oak islands, including a stop at Fort St. Charles. The site is open every day and there is no admission charge.

Fort Scott

FORT SCOTT, KANSAS

THIS WESTERN OUTPOST, established to protect the first trickle of white settlers westward during the Indian warfare days, was named for General Winfield Scott and opened in 1842. It was one of the largest defense establishments on the rivers or the plains during the

formative years of this country.

Within a few years after its founding, Fort Scott was populated with an imposing number of experienced Indian fighters, and troops were being constantly summoned to put down Indian outbreaks against white settlers over wide areas of both Kansas and Missouri. By 1853 the Indian frontier had moved farther westward, and Fort Scott was put on a standby basis and troops withdrawn.

On May 15, 1855, the United States Government sold at auction all the buildings constituting the fort establishment. Headquarters House became a hotel to be named Free State Hotel. On the opposite side of the Plaza a former fort barracks building became the Pro-Slavery Hotel and the Headquarters for a large force of pro-slavery zealots. The antipathy between the anti- and pro-slavery factions was almost visible in the atmosphere. Pro-slavery bandits operating out of Fort Scott drove a large number of white settlers from their land and homes. In retaliation, John Brown, the Free State leader and one of his lieutenants, Captain Bain, established a squatters' court a few miles north of Fort Scott. The stone build-

Officers' quarters, Fort Scott (Photo: Kansas State Historical Society, Topeka)

ing they occupied became known as Fort Bain. They meted out their own justice to pro-slavery zealots until a federal judge, Joseph Williams, sent a posse out to close Fort Bain.

On the other side of the slavery street, the Marais des Cygnes massacre north of Fort Scott in which eleven Free State settlers were lined up in a ravine and killed by pro-slavery activists, was conceived and planned in the Pro-Slavery Hotel in Fort Scott.

Throughout the entire Civil War, Fort Scott was a headquarters for military contingents which participated in many engagements with Confederate forces in Missouri and Kansas. One of the bloodiest battles of the war took place north of Fort Scott when Confederate forces commanded by General Sterling Price were routed by Fort Scott troops at the Battle of Mine Creek. Price was seeking to destroy Fort Scott because it was a strategic Union supply center. One contingent of the defeated Price's

army managed to get within firing range of Fort Scott, but gunfire from the fort repulsed the attack.

Fort Scott was deactivated in 1853 after the Indian lands were safe for settlers. It was reactivated again in 1862 for the Civil War period. Today the Fort Scott historical establishment consists of the Fort Scott Museum housed in Headquarters House, one of the early fort buildings erected in 1842 as living quarters for officers. Headquarters House itself is owned by the city of Fort Scott and operated by the Fort Scott Business and Professional Women's Club. Preserved in it are relics of early frontier life.

The Historic Area of Fort Scott is located on U.S. 69, Bourbon County, in Roadside Park, north of the city of Fort Scott. Call Historic Preservation Association of Bourbon County, in Fort Scott, for hours and admission.

Fort Sill

FORT SILL, OKLAHOMA

FORT SILL has hardly passed into history. It is the headquarters of the U.S. Army Field Artillery School together with many related commands and activities. But back in 1852, Captain Randolph B. Marcy surveyed the land from a shoulder of Medicine Bluff and recommended the site to General Phil Sheridan. "In my humble judgement," he wrote, "a military post established in the vicinity of these mountains, and garrisoned by a force of sufficient strength to command the respect of the Indians, would add (greatly) to the efficiency of the army . . . in western Texas." Sheridan staked out Camp Wichita (which later became Fort Sill) along Medicine Bluff Creek, just north of present Lawton, on January 8, 1869.

The fort was deep in Comanche country, and because the Fort was located where it was, the Indians soon lost the "good medicine" of nearby Medicine Bluff, the three-hundred-foot-high granite and porphyry ridge that young braves used as a retreat for fasting and meditation. Its low stone altar gave way, as a symbol of the area, to the post guardhouse. In time it was housing in bitter defeat such famed Kiowa warriors as Satanta, Satank, Big Tree, and the Apache's Geronimo.

The Indian wars became the first order of military business for the fort, but the fort itself was never attacked. But conditions in the area were far from settled during its first few years. In 1873 and 1874 alone, sixty persons were

Old Army Corral, Fort Sill (Photo: Fred W. Marvel)

Cannons at Fort Sill, Oklahoma, with Geronimo Guardhouse in background (Photo: Fred W. Marvel)

reported killed in the region.

The Indians' attack on a camp of buffalo hunters at Adobe Walls, in the Texas panhandle, was disastrously successful. Fort Sill served as a base for a campaign against the hostile Indians. By mid-1875 the Southern Plains tribes had surrendered and Quanah Parker with his Quahada Comanche was the last to surrender.

With the nearest railroad some three hundred miles away, Sill was forced to be self-sufficient. A sawmill, quarry, and lime kilns were established; soldiers performed most of the chores. Work on the permanent buildings began in 1870. The first structure completed was the stone corral, loopholed for defense. Today, the visitor to Fort Sill will find the historic Old Post located just east of the Main

Post with its modern structures and facilities. In the Old Post almost all of the original stone buildings are still standing and still in use. In addition to the stone corral there are six other exhibit buildings, including the popular Geronimo guardhouse and the graves of Geronimo and other Indian chiefs. The stone corral houses a museum, as does Geronimo's guardhouse. The corral contains military and pioneer horse-drawn vehicles, Apache prisoner-of-war blacksmith equipment, a harness room, a replica of the Old Post Trader's Store, Indian tepees, and other large exhibits. The main floor and basement cells of the guardhouse contain exhibits that commemorate the men and events of Indian Territory days. A highlight of these exhibits is an animated display of Fort Sill as it was in 1875.

An instructive sidetrip from the Old Post to the Fort Sill Museum should be made. It contains the Field Artillery Hall of Flags enshrining the banners of many famous regiments and depicts the history of our national flag, the U. S. Army flag and U. S. Field Artillery flags. Uniforms, paintings, weapons, and equipment from the Revolution to the present accompany the flag displays. The Museum's Cannon Walk that links the Geronimo guardhouse area with the gun halls a block east, features a wide variety of U.S. and foreign weapons from battlefields around the world. The exhibits in McLain Hall depict the history, traditions, and development of the U. S. Field Artillery from the period of the Boxer Rebellion in 1900 to the war in Vietnam. Hamilton Hall houses the story of the American Field Artillery from Colonial times to 1900. The Old Post Chapel built in 1875 is the oldest house of worship in the state of Oklahoma in continuous use since its founding.

Fort Sill is four miles north of Lawton, Oklahoma, on Highways 277, 281, and 62. It is open daily from 9:00 A.M. to 4:30 P.M. and closed on December 25–26 and January 1–2. No admission fee.

Fort Simcoe

OLYMPIA, WASHINGTON

IN EARLY AUGUST of 1856 Colonel George Wright of the 9th Infantry chose the location of Fort Simcoe as he was returning from a pacification march through the Yakima and Kittitas valleys to the Wenatchee River. Indian hostilities had begun in the fall of 1855 with attacks on settlements at the Cascades of the Columbia River. Two forts had to be established to protect the settlers. Fort Simcoe was one and Fort Walla Walla was the other.

The site for Fort Simcoe was a pleasant one. It was spring-fed and had a natural oak grove at the head of Simcoe Valley where the long Toppenish plain meets the foothills of the Cascade Range. It was also the strategic location at the intersection of main trails.

Colonel Wright's expedition was to the salmon fisheries on the Wenatchee River where many Indian bands were concentrated. The main object of the march was to meet up with the Yakima, who disavowed a treaty relinquishing lands to the government. They slew their special agent, Andrew J. Bolon, in the Simcoe Mountains, killed several gold seekers from western Washington, and fought army punitive columns. On October 6, 1855, they attacked Brevet Major Granville O. Haller's command of 104 soldiers just three miles from

*Commandant's House, Fort Simcoe (Photo:
Washington State Department of Commerce and
Economic Development)*

the future site of Fort Simcoe. They inflicted severe casualties and escaped across the Simcoe Mountains. Shortly thereafter, Major Gabriel J. Rains, 4th Infantry, led 700 regulars and volunteers up the Yakima Valley and forced the Indians to abandon the region.

Major Selden Garnett began construction of Fort Simcoe on August 8, 1856, with Companies G and F, 9th Infantry. The first quarters were of hewed pine logs from the hills in back of the post. All lived in tents until the completion of the first barracks in December of 1856.

Renewal of the hostilities began in the sum-mer of 1858 in the Spokane country by the Spokane, Coeur d'Alene, Palouse, and some Yakima. With a large column from Fort Simcoe, Major Garnett moved north to the Okanogan district, capturing, killing, or scattering elements charged with molesting parties of miners. Garnett's column covered 550 miles in thirty-one marching days. When Major Garnett was within a day of reaching Simcoe, he received word that Mrs. Garnett and their son, aged seven months, had just died of "bilious fever." He accompanied their bodies to New York. Captain Archer moved into the Garnett quarters and continued in command until the

Fort Simcoe in 1858 (Photo: Washington State Parks and Recreation Commission)

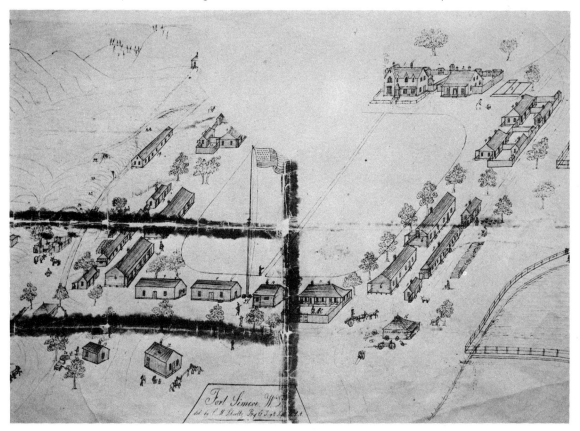

post was transferred to the Indian Department, May 22, 1859.

Of the military structures that framed the 420-foot-square parade ground, only five remain: the commanding officer's house, three dwellings that were captains' quarters, and a squared-log blockhouse on a slight elevation at the southwest approach. Gone are the lieutenants' quarters, barracks, storehouse, subsistence warehouse, guardhouse, hospital, quarters for servants and laundresses, and a little mule-power sawmill.

Walls of the quarters on Officers Row are bricked top to bottom, between the studding, for insulation. The rooms were heated from fireplaces. Bricks were kiln-baked on the post. The blockhouses at the easterly approaches are restorations. The mast-type flagpole is a replica for displaying the thirty-two-star flag flown by the garrison.

Louis Scholl, German émigré of 1848, employed as clerk and draftsman by the quartermaster at Fort Dalles, Oregon, drew the working plans for Fort Simcoe. "All doors, window sashes, mantels, bookcases, etc., were transported by pack mules and large six-mule wagon trains." For two years "a packtrain of nearly fifty mules moved between Fort Dalles and Simcoe."

Fort Simcoe is thirty-eight miles southwest of Yakima, Washington. It is at the western end of State Highway 220. A museum on the grounds contains displays of Indian craftwork and other exhibits depicting the story of Fort Simcoe and the region. The site is open daily all year round. No admission fee.

Fort Sisseton

FORT SISSETON, SOUTH DAKOTA

SOUTH DAKOTA was dotted with forts in the 1800s. The sight and sounds of infantrymen and cavalry horses were a familiar sight to many South Dakotans, especially those living along the Missouri River. Among the many, Fort Sisseton was a major post. It was in 1864 when Major Robert Rose just stumbled upon a natural fortification atop the Coteau des Prairies. Rose and his cavalry troop had combed eastern Dakota Territory looking for a suitable site for a new military installation. A fort was needed to provide military protection of settlers in this new region. Only two years before, some bands of Santee Sioux living along the Minnesota River, a short distance from the fort site, rebelled at unfair trading practices, unkept treaty promises, and settlers homesteading their lands. This uprising, led by Chief Little Crow, raged for five months. There were nearly one thousand casualties. The panic caused by this uprising led to abandoned settlements, and Major Rose was looking for a logical fort site. He found it on the plateau of the Coteau hills. The area was surrounded on all sides by twelve feet of water, and the only possible entrance to the area was a bar that connects two of the lakes. This was the beginning of Fort Sisseton.

It began as Fort Wadsworth and remained as such until 1876, when the War Department discovered they already named a Fort Wadsworth in New York. The name was changed to Sisseton in honor of the Indian tribe that still lives nearby. Because timber was scarce on the plateau, most of the buildings were built of stone or of bricks made by soldiers. Much of the lumber used was transported twenty miles from present-day Sica Hollow. The breastworks of

The stone buildings in northeast South Dakota's
Fort Sisseton (Photo: South Dakota Tourism)

Members of the South Dakota National Guard Artillery Half-Section prepare to march in the Parade Ground at Fort Sisseton (Photo: South Dakota Tourism)

the fort were mounds of sod surrounded by six-foot-deep trenches.

Colonel Henry H. Sibley was put in command of the fort. Sibley was a trader with more than twenty years' experience in the West before being called to military service. The first building he put into service was a log stable for the horses with a roof of branches and slough grass. A mule-powered sawmill was put into operation to help in construction, and in 1866 it was replaced by a twenty-horsepower steam sawmill. The lime and fine clay for making bricks and the abundance of fieldstone for building purposes can still be seen in the beautiful stonework and uniquely colored brick structures that were built to replace the unhewed log facilities.

Part of Sibley's command was a group of Indian scouts under the first chief of scouts, Joseph R. Brown. Shortly after establishing scout headquarters, he built a trading post near the east entrance of the sod breastworks. Probably the most exciting story about Fort Sisseton concerns Brown's son, Sam. In 1866 Fort Sisseton's soldiers had one close call with a battle, but a desperate ride by Scout Brown prevented the disaster. He rode out to warn of an impending Indian raid. When he arrived at his destination, he discovered the Indians were not on the warpath but friendly. He jumped back on his horse and retraced his ride, correcting his error and preventing a battle. At the end of his three-hundred-mile journey through a freezing blizzard, Brown found his legs frozen. He remained paralyzed for the rest of his life.

Fort Sisseton never saw a battle. The fort's soldiers were used mainly as a show of force. After twenty-five years as a military installation, it was abandoned in 1889. Although many of the buildings tumbled down with age, others remained standing and these were restored in the 1930s. The visitor today can enter the North Barracks, originally designed to house two companies of soldiers (about 150 men). It is constructed of split fieldstone and is now the visitor center, lounge and program rooms. The South Barracks is the same size, and the wall mountings where the soldiers' bunks hung can be seen. There is the oil house used to store oil for lamps, machinery, and other needs of the fort. The guardhouse was built with two rooms for the guards and two cells for prisoners. The magazine is directly west of the guardhouse. It is built of stone and brick and was used to store small-arms ammunition, canister shot, and powder for the artillery pieces. The commissary sergeant's quarters are outside the sod breastwork and built of split fieldstone. The building north of the magazine is the adjutant's office. The adjutant is the administrative assistant to the commanding officer and takes care of the bookwork. The officers' quarters have been restored to its original appearance. It was divided into four apartments for officers and their families. The two-story brick building is the commanding officer's residence. The doctor's quarters are next to the commanding officer's residence and the hospital is in front of the doctor's quarters. The library-schoolhouse housed a ninety-four-volume library, a telegraph office, and the post's schoolrooms. The stable is over two hundred feet long and was constructed of split fieldstone. Other restored buildings are the carpenter's, blacksmith's, and wheelwright's shops, the sutler's store (trading post), and the post bakery.

Fort Sisseton is located on South Dakota 73, about twenty miles southeast of Britton, and is now a state park. It is open to the public daily until 11:00 P.M. during the summer and may be viewed in the winter. To enter Fort Sisseton a $3.00 State Park user's fee must be purchased for the visitor's vehicle.

Fort Smith

FORT SMITH, ARKANSAS

IN THE MIDST of a busy city, the remains of two successive frontier forts and the Judge Parker Court stand as reminders of a time when civilization and security ended on the banks of the Arkansas River. At Fort Smith, Indians, lawmen, and outlaws played their dramatic parts in an era that changed the face of the Indian Country. Here, blue-clad troopers gathered to carry the American flag westward, and from here, the men who "rode for Judge Parker" crossed Arkansas to spread the influence of his famous (and some say infamous) court into the lawless land beyond.

The story of Fort Smith falls into three phases: the small first fort, 1817–39, the enlarged second fort, 1838–71, and the Federal District Court, 1872–96. The small first fort was launched when Major William Bradford received orders "to select the best site . . . and thereon . . . erect as expeditiously as circumstances will permit, a stockade." Under these orders he reached Belle Point, a rocky bluff at the juncture of the Poteau and Arkansas rivers on December 25, 1817, and selected it as the

Cold Barracks Building, Fort Smith

site. This confirmed its selection by Major Stephen H. Long earlier. Using Long's plans, Bradford set his sixty-four men to construct a simple wooden stockade with two blockhouses. This unimposing fort was later named Fort Smith for General Thomas A. Smith, commander of the 9th Military Department. Construction proceeded slowly, and not until February 1822 was the fort "nearly completed and in a good state of defense."

The fort's location was chosen to keep peace between the Osage and the Cherokee and to prevent white men from encroaching on Indian lands. Tribesmen of the Cherokee Nation had recently moved into what is now northwestern Arkansas. Penetration into Osage hunting grounds by these Indians from east of the Mississippi produced a constant threat of war. Until 1824, soldiers from Fort Smith were able to prevent serious Indian outbreaks.

By then the frontier had shifted westward, and the fort's garrison moved some eighty miles up the Arkansas River where they established Fort Gibson. Only small detachments returned from time to time to the fort before it was finally closed in 1839. Once deserted, the old fort decayed. Its exact location remained lost until archaeologists uncovered the foundation in 1958.

Demands from the people of western Arkansas for protection against possible Indian uprisings caused Congress to authorize the War Department to reestablish Fort Smith in 1838. Plans called for an impressive installation to be located near the earlier fort. Work began in 1839, but the fort was never completed as planned. By 1841 the danger of uprising had faded. Colonel (later President) Zachary Taylor, the newly appointed departmental commander, opposed further work on the partially completed fort. The government transformed the fort into a supply depot. Occupied by troops in May 1846, the second fort served to equip and provision other forts established deeper in Indian Territory.

Both the North and the South used its supply and hospital facilities during the Civil War. This prolonged its usefulness for a brief time, but the fort's days as a military post were over. By 1871 the Army had little need for such a post at Fort Smith, so the War Department abandoned the reservation.

Today only the barracks and the commissary buildings remain to mark the site of the second fort. All of the walls and other buildings have been pulled down. The stone commissary building, except for minor alterations, appears much as it did when completed in 1846.

The soldiers' barracks, however, has undergone many changes. Completed in 1846, the two-story barracks burned three years later. A second and somewhat smaller barracks was erected on the still intact walls. By 1891 this structure was again enlarged to two full stories to match the new wing added to house the federal jail.

Two decades—the 1870s and 1880s—overshadow the fort's military purpose and represent the period of Fort Smith's greatest fame. In 1872 it was occupied by the U.S. Federal District Court for western Arkansas and the Indian Territory. Between 1875 and 1889 it sheltered one of the most famous tribunals in American legal history, the court of Isaac C. Parker.

There was little law west of St. Louis. The Indians were subject to their own tribal courts but these had no jurisdiction over white men. Criminals and outlaws from all over the United States found sanctuary from arrest or extradition. They came in ever-increasing numbers. Disorder ruled, and reputable men—white and Indian—called upon the federal government for relief. The government assigned jurisdiction over the huge territory to the U.S. District Court for Western Arkansas. It moved to Fort

Would-be offenders possibly had second thoughts after viewing this impressive method of law enforcement at Fort Smith, Arkansas (Photo: Department of Parks & Tourism, Little Rock, Arkansas)

Smith in 1871, and for the next seventeen years the now abandoned soldiers' barracks became the court's permanent home. Its record was very unspectacular. It reached its lowest point just before 1875. To remedy this ineffectiveness, President Ulysses S. Grant appointed Isaac C. Parker to a judgeship vacated by the previous justice, who resigned under threat of impeachment because of a bribery charge. At thirty-six Parker was the youngest member on the federal judicial bench.

Parker approached his task with amazing zeal. For twenty-one years his court at Fort Smith ground out rapid and impartial justice. No appeals from the judge's decisions were possible during the first fourteen years. The volume was astounding. Of some 13,490 cases docketed, 12,000 were criminal in nature. Three hundred and forty-four men stood before Parker accused of major crimes; 160 were convicted. Seventy-nine were hanged. Trials began at 8:30 A.M. and continued into the night. Only Sundays and Christmas halted the routine of the court.

Parker was supported by United States deputy marshals. Dedicated, undeterred by hardship, low pay, or physical danger, they called themselves "the men who rode for Parker." At times as many as two hundred deputy marshals ranged over the court's vast domain. At least sixty-five were killed and numerous others wounded in line of duty. Parker's most famous collaborator was George Maledon, often called the "Prince of Hangmen," who executed sixty of Judge Parker's death sentences. As more and more of the Indian country was opened to settlers, new, local courts whittled away portions of Judge Parker's jurisdiction. Finally in September 1896, his court lost its Oklahoma jurisdiction. Parker was destined to outlive his famous court by only two months. He died at the age of fifty-eight. He looked seventy. The passing of the Parker court followed closely the vanishing of the frontier.

Fort Smith is located on Rogers Avenue between Second and Third streets in downtown Fort Smith. It can be reached from Garrison Avenue (U.S. 64) by turning one block south to Rogers Avenue. The Visitors' Center (old Barracks Building) is open from 8:30 A.M. to 5:00 P.M. daily except December 25 and January 1. No admission fee.

Fort Snelling

SOUTH MINNEAPOLIS, MINNESOTA

ONCE THE MOST northwesterly military post in the United States, Fort Snelling was constructed in the period 1819–23 to protect frontier settlements from Indians and to promote the fur trade. It became the northern outpost of a line of frontier forts—Leavenworth, Gibson, Towson, Smith, Atkinson, and Jesup—guarding the "permanent Indian frontier." Later, from 1861 to 1946, it also served as a training installation.

Just after the end of the War of 1812, the Army—seeking to extend U.S. control over the upper Mississippi Valley—planned a fort at the confluence of the Mississippi and Minnesota rivers on land that Zebulon M. Pike had purchased in 1805 from the Sioux Indians. In 1819 Lieutenant Colonel Henry Leavenworth led a detachment up the Mississippi from Prairie du Chien to build the projected fort and camped for the winter near an Indian village. The following year Colonel Josiah Snelling assumed command and within two years essentially completed the fort. Originally called Fort St. Anthony in 1825, it became known as Fort Snelling.

Fort Snelling guarded the vast region between the Great Lakes and the Missouri River. Few expeditions departed from the fort to attack the Indians, but officials cooperated with Indian agent Lawrence Taliaferro in preventing clashes between the Sioux and Chippewa. Troops from the fort quelled the Winnebago in the Prairie du Chien area, and policed the Canadian border to prevent French-Canadian hunters from crossing it to hunt buffalo. In 1849 troops from Fort Snelling joined dragoons from Fort Gaines to investigate Indian disturbances in Iowa, which later resulted in the founding of Fort Dodge.

After the frontier advanced to the Great Plains, Fort Snelling's importance declined, and in 1857 the Army abandoned it. Reactivated in 1861 as a training center for Civil War troops, it aided in quelling the Sioux uprising of 1862 and from then until 1946 served as a training center. At that time the Army aban-

Round Tower, Fort Snelling (Photo: Department of Economic Development, State of Minnesota)

doned the fort and transferred it to the Veterans Administration, which has released large portions of the military reservation to the state but still owns most of the original fort. Throughout the years other portions of the original reservation have come under private ownership.

Still standing are four of the original sixteen structures; these consist of the quarters of the commanding officer, the officers' quarters, a hexagonal tower, and a round tower. Excavations by the Minnesota Historical Society in 1957–58 uncovered the foundations of several structures, including the powder magazine, schoolhouse, sutler's store, hospital, shops, cistern, and a portion of the original walls.

Fort Snelling is in Hennepin County, at the junction of Minnesota 55 and 100, South Minneapolis, Minnesota. It is open daily, no hours specified. Admission fee not specified.

Fort Spokane

COULEE DAM, WASHINGTON

B Y THE END OF THE 1870's, most Indian bands of the Pacific Northwest had been settled on reservations. This consolidation of tribes together with improved railroad and boat transportation eliminated the need for the many small, stockade-type forts. But the settlers of the upper Columbia had to be protected and a watchful eye had to be kept on the Indian reservations. In 1880 a board of army officers selected a central location for a fort to meet these requirements and they selected a site at the confluence of the Columbia and Spokane rivers.

They transferred troops of the 2nd Infantry from Camp Chelan to construct the fort. During the first year, many temporary buildings were erected by the troop labor with a minimum of funds. After the sum of $40,280.07 was appropriated by the Army for construction of a permanent post, construction was begun in earnest, and during the next twelve years over forty-five buildings were erected.

The Indians were peaceable and the settlers needed little protection, so for the most part, the soldiers enjoyed routine garrison life. Over a twenty-year period, troops of the 2nd, 4th, and 16th Infantry and the 2nd Cavalry saw duty at Fort Spokane. With the outbreak of the Spanish-American War, the entire garrison was withdrawn and equipment and furnishings moved by wagon to the newly established Fort George Wright near Spokane.

Between 1898 and 1929, the former army post was used by the Office of Indian Affairs as headquarters for the Colville Agency, boarding and day schools, and as a tuberculosis sanatorium and general hospital for Indian children.

After 1929 the abandoned fort became a popular picnic site, and through the years, some of the buildings were destroyed by fire and vandalism, while others were sold or moved to the new Colville Indian Agency Headquarters near Nespelem. Fort Spokane was transferred to the National Park Service in 1960, and today, only four of the original buildings remain: the quartermaster stable (1884), the powder magazine (1888), the reservoir (1889), and the guardhouse (1892).

Rows of excavated foundation stones mark the location of three of the six two-story barrack buildings which housed the companies of infantry and cavalry. The barracks contained a kitchen, mess hall, washroom, day room, orderly room, and rooms for the company clerk and two cooks on the ground floor. The second story was a dormitory large enough to hold an infantry company of thirty-three men. The furnishings were spartan. Each man had a bunk with straw-filled mattress, two blankets, and a locker for his clothes and personal items. Company furnishings included wooden chairs and tables, kerosene lamps, and woodburning stoves for heating and cooking. Most of the furniture was made in the post carpenter shop and reflected the rough handicrafts of the frontier. Two bathtubs in each barrack were hardly adequate to keep thirty-three men clean, and an inside water closet was a luxury reserved for officers' homes.

The quartermaster barn is the oldest of the four remaining buildings. Up to fifty-eight mules could be stabled in the barn, with an additional sixteen in an attached shed. The powder magazine was built under civilian contract. Ammunition and powder were secured in the small rear room, while the front room was used for reloading ammunition and repair of arms. The standard infantry arm of the period was the 1873 Springfield rifle. It weighed about eight pounds and was single-shot. The cavalry was issued short-barreled models of the same rifle, called carbines, while officers carried .45-caliber, single-action Colt revolvers. The guardhouse now contains exhibits and artifacts connected with the fort.

Fort Spokane is located in the Coulee Dam National Recreation Area northwest of Spokane. State Highway 155 reaches the area. The fort is open daily from September to June from 8:00 A.M. to 5:00 P.M.; on holidays from 8:00 A.M. to 8:00 P.M. No admission fee.

Sutter's Fort

SACRAMENTO, CALIFORNIA

JOHN SUTTER, born of Swiss parents in 1803 in Kandern, Baden, Germany, sailed to America in 1834, leaving his wife and four children behind not to see them again until sixteen years later in California. Arriving in New York, July 1834, Sutter went on to Missouri, kept store, traded in Santa Fe, and dreamed of wider fields. In 1838 he came with a caravan of fur traders and others to the Pacific Northwest. At Fort Vancouver, Pacific Coast headquarters of the Hudson's Bay Company, he took ship to the Sandwich (Hawaiian) Islands. Later he sailed on a trading vessel to the Russian colony at Sitka, Alaska, and thence to California in July 1839.

The following month Sutter obtained three small vessels in Yerba Buena, as San Francisco was then called, sailed up the Sacramento River, and landed near what is now the head of Twenty-eighth street. On a knoll not far away he subsequently built his fort of sun-dried adobe bricks to guard his territory.

In order to qualify for a land grant, Sutter became a Mexican citizen under Governor Juan B. Alvarado, and agreed to colonize his land and preserve order. Mindful of his former homeland, Sutter called his grant of 48,000 acres New Helvetia. With the help of Indians and the few white men available, Sutter built his fort, raised livestock, and developed agriculture.

During the conquest of California by settlers and armed forces from the "States," the American flag was raised over Sutter's Fort in July 1846. According to a contemporary account, "at this time the Fort is manned by about 50

Sutter's Fort (Photo: State of California, Department of Parks and Recreation)

Sutter's Fort (Photo: State of California, Department of Parks and Recreation)

132

well-disciplined Indians and 10 or 12 white men, all under the pay of the United States. . . . The garrison is under the command of Mr. Kern, the artist of Captain Fremont's exploring expedition."

After the discovery of gold in 1848 Sutter went to Hock Farm, his place on the Feather River near Yuba City, where he was joined in 1850 by his wife, daughter, and three sons. In 1865 his home was destroyed by fire and he returned to Lititz, Pennsylvania. He died in Washington, D.C., in 1880.

Some of the dramatic highlights in Sutter's Fort history are that Kit Carson and an exploring party arrived at Sutter's Fort in 1844 and many emigrants followed thereafter. M. G. Vallejo, commander of the Sonoma garrison under Mexico was taken into custody by Bear Flag rebels and confined in the fort in 1846. U. S. forces were stationed at the fort during the conquest of California from 1846 to 1847. James W. Marshall brought the first gold discovered at Sutter's sawmill to the fort, January 1848. Sutter hosted Colonel Mason, Lieutenant William T. Sherman, Quartermaster Folsom, and other dignitaries en route to inspect the newly discovered mines, July 4, 1848. Sutter's eldest son ordered the first survey of the city of Sacramento in 1848.

Passing out of Sutter's hands shortly after the discovery of gold, Sutter's Fort started to decay. In 1890 the Native Sons of the Golden West arranged for its purchase through public subscription. In 1891 the fort was donated to the state. At that time only the central structure was standing. Reconstruction began almost immediately. Today the visitor can see the Central Building, extensively repaired. Thirty-six original floor joists of whipsawed, hand-hewn oak about fifteen feet long and five by seven inches wide are in their original place. On the upper floor are restored offices of Sutter and his clerks. Sutter's quarters with three connecting rooms representing the kitchen, bedroom, and office of Sutter have been restored, as have the Indian guardrooms that used to house Sutter's personal guard of twelve or fifteen Indians. There is also the blacksmith shop with old tools, forge, bellows, etc., a bastion and dungeon containing cannon and prison and the gunsmith, storage and living quarters, kitchen, meat room, and carpenter shop. The Sutter Story Room contains exhibits and a diorama.

Sutter' Fort is in Sacramento, east of the state capitol, facing Capitol Avenue. It is located at 2701 L Street. It is open from 10:00 A.M. to 5:00 P.M. daily. Adults 50¢.

Fort Tejon

GRAPEVINE CANYON, KERN COUNTY, CALIFORNIA

IN NOVEMBER 1852, Edward F. Beale was appointed commissioner of Indian Affairs in California and Nevada. He established San Sebastian Reservation in Tulare Valley about twenty miles north of the present location of Fort Tejon. In his capacity as commissioner of Indian Affairs for California, General Beale

recommended the establishment of a military post about fifteen miles southwest of the Sebastian Indian Reservation to protect the Indians in the southern San Joaquin Valley and government property at the reservation. The location was considered strategic because an important pass in the Tehachapi Mountain Range

Fort Tejon (Photo: State of California, Department of Parks and Recreation)

could be controlled. Stolen horses and cattle from the San Joaquin were driven through this pass to markets in the Southwest.

Contemporary descriptions of Tejon by visitors include "The post of Tejon is on a little plain entirely surrounded by high mountains, beautifully situated in a grove of old oak. An oasis in the desert where all is freshness and life."

Construction of the post on the fort's present site was authorized June 23, 1854. On June 30 a detachment of Company A, 1st U.S. Dragoons, under the command of First Lieutenant Thomas F. Castor, was ordered to the selected site to begin construction. Fort Tejon later became regimental headquarters for the 1st U.S. Dragoons.

It was an active small post. Patrols traveled as far east as the Colorado River, penetrated unexplored regions of the Owens Valley; rode the supply route to and from Los Angeles; and on occasion escorts from Fort Tejon traveled to Salt Lake City. The troops guarded miners, chased bandits, and generally offered protection to the southern part of the state.

The fort was the military, social, and political center between the San Joaquin area near Visalia and Los Angeles. At the height of its activity there were over twenty buildings and Fort Tejon was one of the largest settlements in Southern California. In 1858 a Butterfield overland mail station was established there which extended from St. Louis to San Francisco.

Fifteen of the officers who served at Fort Tejon during its active period later achieved the rank of general in the Civil War. Eight served with the northern forces and seven with the southern.

After the Army abandoned the fort in 1864, the land became part of General Beale's Tejon Ranch. The buildings on the post were used as residences, stables, and sheds. It was from the Tejon Ranch that, in 1939, the original five-acre gift deed was accepted by the state and Fort Tejon became part of the state park system.

Restoration of the post began in 1949. In 1954 two hundred acres were purchased from the Tejon Ranch Corporation. The barracks building and an officers' quarters were structurally complete in 1957. These two buildings, along with an orderlies' quarters, which has been preserved through the past one hundred years, show the type of structures that were used when the fort was active.

An interesting sidelight was the establishment of a Camel Corps. In 1857 U.S. Secretary of War Jefferson Davis imported twenty-eight camels for transporting supplies to isolated posts in the arid Southwest. Under the direction of Edward F. Beale they were used by a wagon road survey party from Fort Defiance, New Mexico, to Fort Tejon. They performed so well that Beale recommended an expansion of the Camel Corps. With the outbreak of the Civil War and the construction of transcontinental railroads, the camel experiment was discontinued. Varying numbers of camels had been stationed at Fort Tejon from November 1857 until their removal to Los Angeles in June 1861.

Fort Tejon is located in Grapevine Canyon, Kern County, on Highway 99. It is thirty-six miles south of Bakersfield and seventy-seven miles north of Los Angeles near the small community of Lebec. The present highway runs through the original area of the old fort. It is open daily and on holidays from 8:00 A.M. to 5:00 P.M. Admission 25¢.

Fort Totten

DEVILS LAKE, NORTH DAKOTA

THE BEST preserved military post of the Indian War period in the trans-Mississippi West, Fort Totten was established by General Alfred H. Terry on the southeast shore of Devils Lake on July 17, 1867. Named in honor of Major General Joseph G. Totten, late chief engineer of the U.S. Army, it was built as part of a plan to place the Sisseton and Wahpeton Sioux on a reservation and one of the posts to protect the overland route from southern Minnesota to western Montana.

The original fort, designed to be temporary, was built of rough logs and chinked with mud. In 1868 a site was located about eight hundred yards south and selected for a new post. The new fort, built by four companies, was constructed of brick made from clay found on the shore of the lake and from mortar burned in the vicinity.

Following the establishment of the fort, the influx of the Indians to the fort began. In 1868 Major J. N. G. Whistler, post commandant, reported that 681 Indians had received rations there. In 1870 the Indian population at Devils Lake numbered 540. In 1871 the first agent was appointed for the Fort Totten reservation. The civilian population in 1870 at the post, which included employees, laborers, hunters, and others, totaled 174.

Disease and danger were common hazards to both whites and Indians alike in a climate where the temperature in the winter sometimes reached forty below zero or more. During February 1870 the post surgeon reported that two bands of Indians numbering 345 had arrived in bad condition and several had died during the past month for lack of food and the severe cold combined. Occasionally a soldier or civilian became lost in a storm and died. Treatment for frostbite at the post hospital was frequent. Rheumatism, catarrh, bronchitis, consumption, venereal diseases, diarrhea, and dysentery were common. Deaths from accidents were numerous. To prevent scurvy, the post maintained a garden where vegetables were grown for the use of the fort.

Life at the fort had its more pleasant aspects as well. Officers and men alike hunted for sport. Holidays were observed. On the Fourth of July men held foot races and sometimes played baseball. Dances were held for both enlisted men and officers, and the enlisted men organized a variety troupe for the post's entertainment.

During the period of its existence, the average number of men stationed at Fort Totten was 150. During Custer's Little Bighorn Expedition of 1876, two companies of the 7th Cavalry from the post rejoined their regiment, leaving only eighty-three officers and enlisted men at Totten.

Following the Indian Wars of the late 1870s and early 1880s, the need for soldiers to police the reservation diminished. As they organized their own police and established their own courts, the Indians themselves took over more and more the work of the military in enforcing the laws and administering justice on the reservation. On November 18, 1890, the military abandoned Fort Totten and turned the buildings over to the Bureau of Indian Affairs.

For almost seventy years thereafter, the buildings around the parade grounds served as an Indian reservation boarding school where trades were taught the boys and the girls were instructed in household arts. The buildings were declared surplus by the Bureau of Indian Affairs and in January 1960 were transferred to the State Historical Society of North Dakota to be preserved as a historic site.

In Fort Totten Historic Park the only cavalry square of its type still intact is found in the Indian Reservation—home of the Cut Head Sioux and Chippewa Indians. The old fort is undergoing extensive reconstruction and contains an excellent pioneer museum. Nearby is Sully's Hill National Game Reserve with its herds of buffalo, elk, and deer. Picnicking, but no camping, is allowed on the site.

Fort Totten is located near State Highway 57, twelve miles southwest of Devils Lake, North Dakota. Visiting schedule and admission fees unlisted.

Fort Totten (Photo: State Historical Society of North Dakota)

Fort Union

WATROUS, NEW MEXICO

After New Mexico became United States territory, the Army set up its department headquarters and principal supply depot at Fort Marcy in Santa Fe. This arrangement proved unsatisfactory, and in April 1851 Lieutenant Colonel Edwin Sumner was ordered to take command of the defenses of New Mexico and "revise the whole system of defense." One of his first acts was to establish Fort Union, to which he moved the headquarters and depot in August. He got his soldiers closer to the Indians—one of the main reasons federal troops were in New Mexico—and away from the Santa Fe, "that sink of vice and extravagance."

The first of the three forts that ultimately occupied the site consisted of long buildings. For a decade it served as the base for military activities in the area and as a key station on the Santa Fe Trail. It became the principal quartermaster depot of the Southwest, receiving supplies from the East and forwarding them to posts throughout the territory. Dragoons and mounted riflemen fought the Indians who attacked the mountain villages to the north and the desert stretches of the Santa Fe Trail to the east. The nomadic tribes had long fought the Spaniards and the Mexicans. Now they fought the Americans who were overrunning their land and killing all the game.

The Jicarilla Apache struck first and open war broke out in the spring of 1854 when Apache ambushed and almost wiped out a company of dragoons. Lieutenant Colonel Philip St. George Cooke and his troops drove the Apaches into the mountains west of the Rio Grande and on April 8 routed Chief Chacon and his warriors. Many of the Jicarilla took refuge with the Ute, who began raiding Colorado and New Mexico settlements later that year.

Colonel Thomas Fauntleroy with five hundred men and a large contingent of volunteers rode north in February of 1855 and after minor skirmishing attacked a Ute camp on April 28, killing forty Indians and burning their lodges. Ute resistance collapsed. Less successful was the campaign of 1860 against the Kiowa and Comanche who raided the eastern borders of New Mexico. After five months of marching, the soldiers returned to Fort Union without firing a shot at an Indian.

In April of 1861 South Carolinians fired upon Fort Sumter and the problem with the Indians became less important. Confederate invasion of New Mexico was expected momentarily and Colonel Edward S. Canby, commanding federal troops in the territory, ordered construction of an earthwork fortification designed for defense. The ditches, parapets, and bombproofs of the second Fort Union were completed late in 1861.

Brigadier General Henry H. Sibley, who had commanded Fort Union before the war, had resigned from the U.S. Army and gone to Texas to raise a brigade of mounted riflemen for the Confederate offensive. By January 1862 he had concentrated some 2,500 men at El Paso for the march up the Rio Grande toward Fort Union's supplies and Colorado's gold fields. Colonel Canby and his men met the advancing Texans on the middle stretches of the Rio Grande, brushed them aside, and pushed on to Albuquerque and Santa Fe. Only Fort Union lay between them and Denver.

News of the invasion reached Colorado, and a hastily raised regiment of volunteers under Colonel John P. Slough decided to attack first.

Military prison, post, and third fort, Fort Union National Monument (Photo: New Mexico National Park Service)

He led his men toward Santa Fe, and on March 26 the advance guards of the two armies clashed in the narrows of Apache Canyon fifteen miles east of Santa Fe. The Confederates were thrown back. Two days later the main armies clashed at Pigeon's Ranch in Glorieta Pass. After a seesaw engagement the Confederates withdrew and New Mexico was securely in Union hands. Work got under way on the third and final fort. Plans provided for a depot with warehouses, corrals, shops, offices, and quarters. The supply installation overshadowed the adjacent post and housed far more men, largely civilian employees.

After the Civil War Fort Union was active in Indian campaigns. Kit Carson led units from Fort Union against the Mescalero Apache, the Navajo, the Kiowa, and the Comanche. In the winter of 1868 Major General Philip H. Sheridan organized a campaign against the Cheyenne, Arapaho, Kiowa, and Comanche of the Southern Plains. Although his operations in 1868-1869 brought the tribes to terms, they revolted again in 1874. Once again Fort Union and other New Mexico posts furnished several columns of troops that skirmished with the Indians throughout the fall, winter, and spring of 1874–75.

The end of the Indian Wars and the arrival

of the Santa Fe Railroad were the death knell of Fort Union. It had ceased to be of much use and was abandoned in February 1891. In 1956 Fort Union became a national monument, and after being abandoned for some eighty years the problem was how to preserve its ruins. It was decided not to reconstruct the fort. It was decided to preserve the ruins and not to let them fall further into disrepair. All chimneys rest on and are filled with reinforced concrete. Missing rocks in the foundations have been replaced and walls repainted. Adobe walls have been sprayed with silicone, which makes them water-resistant.

The visitor need not follow any set trail or sequence in viewing the fort. The fort consisted of the post, the depot, and the arsenal across the valley to the southwest. No stockade or wall surrounded the fort. A walk about its ruins will give some small indication of its size and importance of this once-major post.

Fort Union is eight miles north of I-25 at the end of New Mexico 477. Watrous, New Mexico is a half mile south of the intersection of these two highways. The nearest large community is Las Vegas, twenty-six miles to the south. The park and fort are open during daylight hours only and every day except January 1 and December 25. No admission fee.

Fort Vancouver

VANCOUVER, WASHINGTON

For two decades this stockaded fur-trading post was headquarters and depot for all activities of the Hudson's Bay Company west of the Rocky Mountains. As such, it was the economical, social, political, and cultural hub of an area now comprising British Columbia, Washington, Oregon, Idaho, and western Montana.

The fur resources of the Pacific Northwest were discovered by British seamen who visited the northwest coast and obtained valuable furs in trade with the Indians about the time of the American Revolution. Soon traders from several European countries, Canada, and the infant United States were competing for the riches thus revealed. After years of bitter con-

Fort Vancouver, 1845

test, the Hudson's Bay Company, a British firm chartered in 1670, won a dominant position in the northwest fur trade.

In 1824 the company decided to move its western headquarters from Fort George, at the mouth of the Columbia River, to a site about one hundred miles upstream, to where the State School for the Deaf in Vancouver, Washington, now stands. This shift was made to strengthen British claims to the territory north of the Columbia. The new post was named Fort Vancouver in honor of Captain George Vancouver, the explorer.

In 1829 a new fort was built one mile west and closer to the river. From that time the post grew in size and importance. Between 1824 and 1846 Fort Vancouver was commanded by Chief Factor John McLoughlin. Of towering height, he was known to the Indians as "Whitehead Eagle." Under his leadership, the Hudson's Bay Company won a virtual monopoly of the fur trade in the Oregon country. Fort Vancouver was the nerve center of a vast commercial empire. From its warehouses went out supplies for the many interior posts, for the fur brigades which ranged as far distant as present-day Utah and California, and for the vessels and forts which dominated the coastal trade well up to the shoreline of Alaska. At Fort Vancouver each year the fur returns for the entire western trade were gathered for shipment to England.

The fort was also the center for an important farming and manufacturing community. The fort's cultivated fields and pasture lands extended for miles along the north bank of the Columbia. Lumber, pickled salmon, and other products of Fort Vancouver's mills, drying sheds, forges, and shops supplied not only the wants of the fur trade but also a brisk trade with such distant places as the Hawaiian Islands, California, and the Russian settlements in Alaska. Fort Vancouver marked the begin-

ning of large-scale agriculture and industry in the Pacific Northwest.

At the height of its prosperity—about 1844 to 1846—Fort Vancouver was an extensive establishment. The fort proper measured about 732 feet by 325 feet. It was surrounded by a stockade of upright logs; a bastion at the northwest corner mounted seven or eight thirty-pound guns. Within the stockade were about twenty-two major buildings and several lesser structures. Among the former were four large storehouses, an Indian trade shop, a granary, an impressive residence for the chief factor, dwellings for other company officials and clerks, and a jail.

The lesser employees—the tradesmen, artisans, boatmen, and laborers—had their homes in what was known as "the village," on the plain west and southwest of the stockade. It consisted of from thirty to fifty wooden dwellings. Near the village and extending to the river was a lagoon, around which were a number of other company buildings including a wharf, a storehouse for salmon, boat sheds, and a hospital.

The treaty of 1846 between the United States and Great Britain established the 49th parallel as the southern boundary of Canada. Fort Vancouver thus found itself in American territory. Thereafter the influence of the post and the Hudson's Bay Company declined rapidly south of the Canadian line. Settlers began to take over land near Fort Vancouver, and to protect itself, the company welcomed the establishment of a United States Army camp in 1849. A short time later, a military reservation was created around the old fur-trading post. The last factor of Fort Vancouver handed over the keys to the army quartermaster in 1860. Six years later it was reported that all traces of the old stockade had been destroyed by fire of unknown origin.

The army post at Fort Vancouver continues

active to this date. Known variously as Camp Vancouver, Fort Vancouver, and Vancouver Barracks, it long served as military headquarters and supply point for the Pacific Northwest. Though now much reduced in size, Vancouver Barracks commemorates the role of the United States Army in the settlement and development of the West. Among the many men who served at the Vancouver army post were Grant, Custer, Sheridan, and George C. Marshall.

Fort Vancouver is located in Vancouver, Washington, between Vancouver Freeway, East Mill Plain Boulevard, and the Columbia River on the south. State Highway 99 can be used to reach the post. It is open Monday to Friday from 8:00 A.M. to 5:00 P.M.; Saturday, Sunday, and holidays from 9:00 A.M. to 5:30 P.M. No admission fee.

Fort Vasquez

PLATTEVILLE, COLORADO

AT FIRST, the Far West attracted the "daring men"—the "mountain men"—who not only explored the routes that would open the West to eventual settlement but started the fur trade, which reached a peak in the 1820s. In the next decade when the beaver had been trapped almost to extinction, a new fur-trade era was opened. It was the era of the traders whose annual rendezvous with the mountain men and the Indians was part shrewd business and part carouse. These open-air markets were superseded by a series of trading posts scattered through the Rockies. Fort Vasquez was such a post.

Two experienced mountain men, Louis Vasquez and Andrew Sublette, built Fort Vasquez about 1835. It was intended to capture the trade of the Indians along the South Platte. It was described by a contemporary as being about one hundred feet square with walls about twelve feet high. The walls, built on a sandstone foundation, were pierced at intervals by rifle ports, and towers rose above two of the corners. Mud for the adobe bricks was prepared by driving oxen around to tramp and mix it. Inside the fort were living quarters, a barn, storage rooms, and a trading area. It was a substantial structure.

The South Platte must have proved to be a profitable one for trade because competing posts were quickly erected in the vicinity. Former army officer Lancaster P. Lupton built Fort Lupton a few miles to the south in 1836 or 1837. Peter Sarpy and Henry Fraeb built Fort Jackson nearby in 1837. Fort St. Vrain was opened in the same year by the owners of Bent's Fort, the famed "adobe castle" on the Arkansas River. Even distant Fort Laramie felt the competition of traders at Fort Vasquez, which drew the Indians away from the Laramie area.

Vasquez and Sublette hauled their goods to the fort in wagons drawn by mules or oxen on the rudimentary roads that had penetrated the Colorado region even that early. The stock of merchandise at the fort included such items as black silk handkerchiefs, ivory combs, brass tacks, the "best Wilkinson Scalping Knives," Hudson Bay blankets, brass kettles, mule shoes, and, of course, "Taos Lightning." Drunkenness was a common condition at the fort and the alcohol made the Indians more amenable to the high prices charged for trade goods. The Indians usually offered buffalo robes, but beaver skins and even gold dust valued at $2.50 per ounce were also offered.

Fort Vasquez, Platteville, Colorado

144

An attempt was made on at least one occasion to carry the furs from Fort Vasquez to St. Louis on the South Platte in a flat-bottomed Mackinaw boat. Loading on seven hundred buffalo robes and four hundred buffalo tongues, the crew waded and pushed the boat for the first three hundred miles. It took them sixty-nine days to arrive at St. Louis.

By this time competition from other posts forced Vasquez and Sublette to sell their fort. They sold it in 1840 or 1841 to Lock and Randolph, who were even less successful. Poor management, trouble with the Indians, and just plain hard luck forced them to abandon the fort in 1842, leaving Vasquez and Sublette with an unpaid note for eight hundred dollars. The fort soon fell into ruin. Portions of the walls survived until the 1930s, when the fort was reconstructed as a WPA project. At first owned by Weld County and maintained by the citizens of Platteville, Fort Vasquez was deeded to the State Historical Society in 1948. The Fort Vasquez Visitors' Center was completed early in 1964. Exhibits in the fort and the Visitors' Center tell the story of the Colorado fur trade.

Fort Vasquez is located on U.S. 85 one mile south of Platteville, thirty miles north of Denver. It is open daily from 9:00 to 5:00. No admission fee.

Fort Walla Walla

WALLA WALLA, WASHINGTON

FORT WALLA WALLA, parent of the city of the same name, was one of the two regular army posts established in the interior of Washington Territory to guard the surrounding settlements and isolated settlers against Indian raids. No attack was ever made by Indians against the post itself, but time and time again the troops from the fort rode out to protect the settlers.

The first demand for a government military post was when the Presbyterian mission of Dr. Marcus Whitman located on the Walla Walla River was attacked by the Cayuse Indians in 1847 and all the residents (including Dr. Whitman), save six, were massacred. The news of this raid together with others that followed throughout the Northwest caused the authorities to send Colonel Cornelius Gilliam and his troops from Oregon to protect the country. Colonel Gilliam was badly defeated by Indians and withdrew, leaving the country abandoned.

The discovery of gold in the Colville country in 1855 and the subsequent gold rush convinced the politicians that peace was essential and they succeeded in signing treaties with seventeen tribes including the Walla Walla, the Cayuse, and the Umatilla. They persuaded the tribes to give up their lands for the sum of $100,000 cash and an annuity of $500 for twenty years. This treaty didn't last long enough for the signatures to dry on the document. The Indians were dissatisfied with the treaty and the Indian War of 1855–56 was the result. Murders and massacres followed. Calls for help brought volunteer troops from Oregon. They defeated the Indians in a four-day battle in December of 1855. The following spring a regiment of volunteers defeated the Indians in several fights. The first regulars came that autumn. Four companies of the 1st Dragoons and a detachment of the 3rd Artillery under Lieutenant Colonel E. J. Steptoe were sent to council with the Indians and relieve the volunteers. The

Fort Walla Walla, 1857

The Blockhouse, Fort Walla Walla

council lasted several days and ended in a battle with the Indians. In November Colonel George Wright of the 9th Infantry came back with Colonel Steptoe to Walla Walla and established Fort Walla Walla at what is now the intersection of First and Main streets in Walla Walla. Barracks were begun immediately and finished Christmas Day.

In the late spring of 1858 the Spokane, Couer d'Alene, Palouse, and Yakima renewed their raids, and in May of 1858 Colonel Steptoe and his troops set out to contain and defeat the warring Indians. It was a small force and they marched from Fort Walla Walla to the Colville region. The troops were completely routed and most of the force, including Colonel Steptoe, were killed. Colonel Wright, in August, set out to avenge Steptoe and succeeded in defeating the tribes decisively. From that time on there was no trouble with the Indians in the vicinity of Fort Walla Walla.

In the fall of 1858 the country was open to settlers and the city of Walla Walla was started about the fort. Since that time the history of the city and of the fort have been inseparably intermingled. In 1865 all the troops save a small detachment were withdrawn and by an act of Congress the fort was abandoned. In 1873 it was again announced as a military reservation, and although it had been previously abandoned to the Interior Department it was again turned over to the War Department.

From 1873 until June 30, 1884, the 1st Cavalry, together with other regiments, occupied the post. Following them came the 2nd Cavalry, parts of which stayed until June 4, 1890. The 4th Cavalry then succeeded, remaining until August 26, 1905. In December 1905 the 14th Cavalry were sent there, remaining until 1908, when the 1st Cavalry returned to stay until the post was abandoned in 1910. In 1921 the President signed a bill authorizing the establishment of the Walla Walla Veterans Hospital. Several of the original buildings were torn down and some were reconstructed. Seven of the original buildings were left. In 1974 Fort Walla Walla became a museum complex including a pioneer village, a historical museum, the old military cemetery, and a number of early farm implements scattered about the site.

The pioneer village as well as the military cemetery is located on ground that was part of the original Fort Walla Walla. A reconstructed blockhouse is a replica of the original fort blockhouse. The visitor to the fort site can visit some ten historic buildings that are now part of Pioneer Village.

Fort Walla Walla is located on First and Main streets in the city of Walla Walla, Washington. It is open from 1:00 to 5:00 P.M. from May to October on Saturday and Sundays only. It is closed from September to May. Adults $1.00. Children 50¢.

Fort Washita

DURANT, OKLAHOMA

GENERAL ZACHERY TAYLOR founded Fort Washita in 1842 when he was in command of the 1st U.S. Infantry. Six years later he became the twelfth President of the United States. The fort was meant to protect the Chickasaw Indians from the depredations of the wild Plains Tribes and marauding Texans.

During the gold rush of '49 Fort Washita was an important stop on the California Road.

The Texas Road and the Pawnee Trail also crossed here bringing settlers, soldiers, and frontiersmen to the fort. Texas was frontier country and was harassed by the dread Comanche, and only large parties, well armed and well equipped, undertook the long journey into that region. Emigrants remained at Fort Washita until a large wagon train could form, then, with elected leaders, adequate supplies, guides, and scouts, left the fort as one large party.

By the early 1850s the post had deteriorated considerably and it was substantially rebuilt and improved in 1855–56. It was abandoned on April 16, 1861, and occupied by Confederate troops from Texas on the following day. Toward the close of the Civil War, the headquarters for the Indian Territory Department of the Confederacy was moved to Fort Washita. At the close of the war on July 1, 1870, the fort was burned by retreating soldiers, or by settlers who feared that its buildings would become strongholds for bands of outlaws. The post was never again occupied by the U.S. Army. On July 1, 1870, the military reservation, which had never been formally declared, was turned over to the Interior Department for the use of the Chickasaw Nation. When tribal holdings were later allotted and opened for settlement, land containing the fallen-down fort became the property of Charles Colbert. In March of 1962 the Fort Washita complex was purchased by the Oklahoma Historical Society and restoration of the fort began. Visitors can view the ruins of the fort, the reconstructed south barracks, and General Cooper's cabin.

Fort Washita is on Star Route 213, eleven miles east of Madill on Highway 199. It is twelve miles northwest of Durant. It is open Tuesday to Friday from 9:00 A.M. to 5:00 P.M.; Saturday and Sunday from 2:00 to 5:00 P.M.; closed on Monday. No admission fee.

Fort Washita (Photo: Fred W. Marvel)

South

Fort Caroline

JACKSONVILLE, FLORIDA

WHEN FORT CAROLINE was founded, no other European colony existed on the North American continent this side of Mexico. By planting this colony, France hoped to acquire a share of the New World claimed by Spain. The French move forced Spain to act and brought on the first decisive conflict between Europeans for a region that later became part of the continental United States. At Fort Caroline the battle between France and Spain for supremacy in North America was joined.

During this period France was often in trouble exhausted by her European wars. The Admiral of France, Gaspard de Coligny sought to strengthen his country in the New World, and French bases in Spanish America were part of his plan. On June 25, 1564, his expedition of three vessels and some three hundred people anchored off the St. Johns River in Florida. For the site of the colony they chose a broad, flat knoll of the river shore about five miles from its mouth. With Indian help they raised a triangular fort of earth and wood which enclosed several palm-thatched buildings. Other houses were built in the meadow outside the fort, and the colony was named Fort Caroline in honor of King Charles IX.

The new settlement lay in the Timucua Indian country and Chief Saturiba presented a wedge of silver to the commander of the fort, René de Laudonnière, which he said came from hostile Indians farther up the St. Johns. Laudonnière sent envoys upriver; they procured a few more pounds of silver, along with stories of a great chief named Outina, whose allies wore armor of gold and silver. Laudonnière's efforts to promote peace between Outina and Saturiba only alienated Saturiba. There were other troubles. When Laudonnière refused any large-scale exploration for gold until the fort was strengthened against attack, mutineers stole a vessel and sailed southward. After taking a Spanish treasure ship and plundering a Cuban hamlet they were seized by the Spaniards and now Spain had firsthand information about the Florida colony.

During the winter and spring of 1564–65 the Indians withdrew to the forests and the French were close to famine. In desperation Laudonnière seized Outina to ransom him for corn and beans from native storehouses. The exchange was made but as the French left Outina's village, they walked into an ambush and most of their loot was lost. They decided to go back to France. They traded cannon and powder to the English slave trader John Hawkins in exchange for one of his ships, and by August 15 they were ready to leave.

The Spanish considered Fort Caroline a threat to Spanish commerce and feared for their treasure fleets following the Gulf Stream and sailing past Fort Caroline. An armada left Cádiz for Florida in June 1565 to eliminate Fort Caroline. The French, knowing about the armada, sent a fleet under Jean Ribaut with reinforcements. Ribaut reached Fort Caroline on August 28 just as the colonists were about to sail for France. Cargoes went back into the storehouses and there was no more talk of leaving. That same day the armada, under Menéndez, was off the coast searching for the Frenchmen. A few days later he found the French ships anchored at the mouth of the St. Johns. He tried to board them, but they cut the anchor cables and escaped. Menéndez dropped down the coast a few miles and on September 8 established the colony destined to live through the years as St. Augustine.

Fort Caroline Replica

Against the advice of his captains and Laudonnière, Ribault decided to attack the Spanish. In the hurricane season it was a fateful mistake. When a storm blew up, the French fleet was driven ashore and wrecked many miles south of St. Augustine. Menendez knew that Ribault's fleet was gone and that most of the French fighting men were aboard those wrecked ships. He decided that now was the time to attack Fort Caroline. With 500 men, guided by Indians and a French prisoner, he marched toward Fort Caroline. About 240 people were left at the French fort. At dawn the Spaniard swept down upon the fort and they killed 142 of the men and captured about 50 women and children. It was September 20, 1565. After posting a garrison at the fort Menéndez returned to St. Augustine. Ribaut's shipwrecked men suffered a similar fate to those left behind. Helpless and hungry, 350 surrendered to Menéndez. He killed 334 of them. The massacre still bears the name of Matanzas (slaughters).

Destruction of the colony caused a furor in France. A Frenchman named Dominique de Gourgues sailed from Bordeaux with three vessels and 180 men, seemingly equipped for the slave trade, but secretly determined to avenge his compatriots. He landed north of the St. Johns and enlisted Indian allies. Two batteries near the river mouth were captured and the forces moved on to Fort Caroline, now renamed San Mateo. Its guns opened fire. The Spanish made a sortie that was quickly cut down and the garrison fled to the forest—where the Indians were waiting. A bare handful of the Spaniards won their way through to St. Augustine. San Mateo was burned.

The site of Fort Caroline no longer exists. It was washed away in 1880 when the river channel was deepened. The fort walls have been reconstructed upon a vestige of a river plain. The reconstruction was based on a sixteenth-century sketch by Jacques Le Moyne, the colony's artist and mapmaker.

Fort Caroline is about ten miles east of Jacksonville and five miles west of Mayport. It can be reached by Florida 10; turn off on the St. Johns Bluff Road or Girvin Road and then proceed east on Fort Caroline Road. Open daily from 8:00 A.M. to 5:00 P.M. No admission fee.

Castillo De San Marcos

ST. AUGUSTINE, FLORIDA

THE FIRST CONFLICT between Europeans in North America occurred more than four hundred years ago on a fifty-mile stretch of Florida's east coast. Spain claimed this strategic coastland on the basis of discovery. France challenged that claim and alleged earlier exploration, and England, a latecomer, just bided her time until her strength was developed to the point of taking what she wanted.

France was the first nation to establish a foothold in the Florida wilderness that Spain had unsuccessfully attempted to settle several times. The French made an effort to control this region in 1564 when troops under René de Laudonnière built the sod-and-timber Fort Caroline five miles from the mouth of the St. Johns River. Mutiny, hunger, and Indian troubles plagued the settlement and it barely survived.

Despite these problems and the settlement's

Castillo de San Marcos

shaky existence, the mere presence of the fort mocked Spain's claims to Florida and threatened the passage of Spanish treasure fleets that followed the Gulf Stream and sailed close inshore. Spain responded by sending an expedition both to settle Florida and to drive out the French. When the Spaniards, commanded by Pedro Menéndez de Avilés, arrived at the mouth of the St. Johns River in 1565, they tried unsuccessfully to board the French ships. They then sailed to a harbor farther south and established St. Augustine as a base for further operations.

Almost immediately the French, led by Jean Ribaut, sailed south to attack. Their fleet, however, was driven off St. Augustine by a violent storm, and the mission failed. Taking advantage of the French preoccupation off St. Augustine, and realizing that Fort Caroline would be lightly guarded, the Spaniards marched north and attacked the fort. They captured it and executed most of the inhabitants. The French fleet was also in trouble. Forced ashore by the storm many miles below St. Augustine, the survivors began an overland march to Fort Caroline. Learning from the Indians that the French were headed north, the Spaniards moved to intercept them. At an inlet, fourteen miles south of St. Augustine, the antagonists met. Some Frenchmen escaped, but most surrendered and were put to death. The Spanish soldiers maintained that this was a military necessity. The episode gave a name to the area: Matanzas, Spanish for "slaughters."

Spain was not to enjoy unhindered possession of Florida. In 1568 an expedition of vengeful French freebooters attacked Fort Caroline (then called Fort San Mateo), burned it, and hanged the survivors. They took revenge on the crews of captured Spanish vessels by throwing them into the sea. Then in 1586 England flexed its seapower when Sir Francis Drake attacked and destroyed St. Augustine. The townspeople began immediately to rebuild it.

Early in the seventeenth century, England entered North America in earnest to seize Spanish-claimed territory. In 1607 Englishmen settled at Jamestown; by 1653 they had pushed south to settle in present-day North Carolina. The British again sacked St. Augustine in 1668, and this hit-and-run attack, followed by the English settlement of Charlestown (in today's South Carolina) in 1670, caused Spain to build a defensive stone fort at St. Augustine— Castillo de San Marcos. Construction began in 1672 and continued at intervals until 1695.

After the destruction of the French at Matanzas Inlet, the Spanish built a watchtower (Fort Matanzas) at its mouth to warn St. Augustine of vessels approaching the city. Despite this precaution, pirates surprised the Matanzas garrison in 1683 and marched toward St. Augustine and the unfinished Castillo. A Spanish soldier, escaping from the pirates, warned the garrison, which ambushed the pirates and turned them back. In 1686 the Spaniards repulsed another raid, this time at Little Matanzas Inlet, one and a half miles south of the main channel.

Castillo de San Marcos came under fire in 1702 during Queen Anne's War, when the English seized St. Augustine and unsuccessfully besieged the fort. The fifty-day military operation ended with the burning of the city, but Castillo emerged unscathed, thus becoming the symbolic link between the old St. Augustine of 1565 and the new city that rose from its ashes. As disputes with England continued, earthwork defense lines were built on the north and west limits of the settlement, and St. Augustine became a walled city. When English settlers and soldiers moved into Georgia, Spain began to modernize the Castillo. Matanzas, however, was still unfortified when the English struck in 1740. They again laid siege to the Castillo. After a twenty-seven-day nerve-shattering

British bombardment of the fort and the city, the British blundered and prematurely lifted their blockade. This gave the Spanish a chance to bring in critically needed provisions through Matanzas Inlet. Finally, with the hazards presented by the hurricane season, the British fleet sailed away and the army had no choice but to abandon the siege. This abortive attack convinced the Spaniards of the need for a strong outpost. Consequently in 1742 they completed the stone tower at Matanzas and also strengthened the St. Augustine fortifications by modernizing the Castillo and by constructing two additional earthwork lines to the north of it.

In 1763 when Spain ceded Florida to Great Britain in return for British-occupied Havana, the British also strengthened the Castillo, holding it through the American Revolution. For a while, military operations against Georgia and South Carolina originated in St. Augustine,

and three signers of the Declaration of Independence were detained here. By the provisions of the Treaty of Paris of 1783, Great Britain returned Florida to Spain. After the American Revolution, separatists, Indians, and renegades created incidents which led to serious Spanish-American tensions, causing Spain to cede Florida to the United States in 1821. The Castillo, which became known as Fort Marion, housed Indian prisoners during the Seminole War of the 1830s; Confederate troops occupied it briefly during the Civil War; and later on, western Indian prisoners were held there. It was last used during the Spanish-American War as a military prison.

The Castillo de San Marcos is in the city of St. Augustine and can be reached by U.S. 1 and Florida A1A. It is open daily from 8:30 A.M. to 5:30 P.M.; closed Christmas. Adults 50¢. Children under sixteen free.

Fort Clinch

FERNANDINA BEACH, FLORIDA

EIGHT FLAGS have flown over Fort Clinch; France, Spain, Great Britain, Patriot's Flag, Green Cross of Florida, Mexico, Confederate, and it is still flying the flag of the United States.

Fort Clinch was one of the chain of masonry forts on the Atlantic Coast. It was built by the United States on the northern tip of Amelia Island. This island, discovered in 1562 by the French explorer Jean Ribaut, has played a colorful role in our nation's history.

The fort was named for Brevet Brigadier General Duncan Lamont Clinch, an important figure in Florida's Seminole Wars of the 1830s. Construction began in 1847, with its primary purpose to guard passages through Cumberland Sound into the deep-water harbor of Fer-

nandina.

Construction was far from complete when the Confederates seized it in 1861. It was hastily evacuated in 1862 when a combined federal naval and army attack threatened. Federal forces possessed Amelia Island during the remainder of the Civil War.

It was again strengthened and used during the Spanish-American War, and found limited use during World War II as a communications and security post. Fort Clinch was acquired by the state in 1936 for development as a state park. The architecture of the fort is outstanding and is still in an excellent state of preservation.

Fort Clinch is three miles from Fernandina Beach on Florida Road A1A. A standard fee is

charged for all Florida state parks and there are facilities for tent and trailer camping. Open daily 9:00 A.M. to 5:00 P.M. No admission fee to fort.

Fort Delaware

PEA PATCH ISLAND, DELAWARE CITY, DELAWARE

THE "ANDERSONVILLE of the North" was the description given to the imposing and awesome structure situated on Pea Patch Island, in the Delaware River, just one mile from Delaware City. The 178-acre island was named from a colonial-day tale that a boat loaded with peas ran aground and sprouted in sandy loam.

The first fort on Pea Patch in 1813 was an earthwork that was later dismantled, and in 1821 a masonry fort was constructed. It served until 1832 when it was destroyed by fire. In 1847 Congress passed an appropriation of one million dollars to construct the largest modern fort in the country. The state of New Jersey claimed the island but the courts settled the claim in Delaware's favor opening the way to begin construction on the fort (as it is today) in 1848.

Just driving the pilings exhausted the million-dollar appropriation and Congress had to provide another million to continue the work. It was 1849 before actual construction on the fort—which was to surpass Fort Sumter in size—began, and it was not completed until 1859—just two years before the Civil War.

The fort was first occupied by one company of regular artillery in February 1861. The Commonwealth Artillery of Pennsylvania were the first volunteers to move in after the beginning of the war.

After the Battle of Kernstown in 1862 some 250 men of Stonewall Jackson's army—mostly Virginians—were brought to the island as the first Confederate prisoners of war. It had not been planned for such use up to that time. The barracks space was crowded by these prisoners and commanding officer Captain A. A. Gibson complained. Wooden barracks were then erected in 1862 to house some 2,000 prisoners. By June of 1863 there were some 8,000 prisoners on the island and the barracks had been expanded to house 10,000. All of the prisoners captured at Gettysburg from General James A. Archer down to the last private were held at Fort Delaware after the battle. There were 12,500 prisoners on the island in August 1863—an appalling number! About 2,700 prisoners died while incarcerated at Fort Delaware. Some 2,400 are buried in a national cemetery at Finn's Point, New Jersey, just across the Delaware River—adjoining Fort Mott.

Among political prisoners housed at the fort were Burton H. Harrison, private secretary to Jefferson Davis, and Governor F. R. Lubbock of Texas, who was the last prisoner at the fort in 1865.

In 1896, on the eve of the Spanish-American War, after the fort had been in little use for nearly twenty years, Congress appropriated $600,000 to install modern disappearing guns. The fort was again fully garrisoned. In 1903 only a token force remained at the fort, and in 1917 the pattern was again repeated. At the outbreak of World War II another company was moved to the island, and in 1943 the big disappearing guns were removed for scrap. The

*Fort Delaware today (Photo: W. Emerson Wilson
News Journal Co.)*

fort was closed up entirely in 1944 and later turned over to the state of Delaware and is now under the jurisdiction of the Delaware Park Commission. Interestingly enough, this bastion, which was built primarily to protect Philadelphia and its harbor, never fired a combat shot throughout its history.

The walls of pentagon-shaped Fort Delaware are of solid granite blocks and vary in width from seven feet to thirty feet thick and are thirty-two feet high. The fort is surrounded by a thirty-foot moat, crossed by a drawbridge on the Delaware side leading from the sally port—or principal entrance. There are three tiers for guns, two consisting of casemates containing examples of some of the finest brick masonry in the country. This is still in evi-

dence. More than 25 million bricks are said to have been used for the fort and barracks. The circular granite stairways are unique architectural features. There are two barrack buildings facing the parade grounds. Offices of the commanding general and quarters for officers were in the building on the north side. The men, mess halls, and kitchens were located in the barracks on the west side, where the museum is now to be found on the ground floor.

Visitors can reach Fort Delaware on Route 9 to Delaware City, where boats leave the harbor for the fort every hour from noon to 6:00 P.M. It is open every Saturday, Sunday, and holiday between May and mid-September. Transportation charges: adults $2.00, children $1.00.

Fort Delaware (Photo: W. Emerson Wilson News Journal Co.)

Fort Donelson

DOVER, TENNESSEE

FORT DONELSON was the scene of one of the early decisive battles of the Civil War. A victory for the Union forces under the command of Brigadier General Ulysses S. Grant, the surrender of Fort Donelson and some 12,000 to 15,000 Confederate officers and men first directed wide public attention to Grant as a military leader of high caliber. Also, after almost a year of war in which the Confederates had been uniformly successful, this victory did much to raise the flagging spirits of the supporters of the Union cause.

In his campaign against Fort Donelson, a typical earthen Civil War field fortification, General Grant, for the first time in the war, made successful use of a river for large-scale operations. His attack, brilliantly conceived and executed, resulted in the most important victory yet achieved by the North. It opened an avenue into the very heart of the Confederacy by way of the Tennessee and Cumberland rivers, forcing the immediate evacuation of Columbus and Bowling Green, and delivering western Tennessee and all of Kentucky into federal hands. The battle marked the beginning of a campaign which, after seventeen months of bloody fighting, resulted in the complete control of all strategic points in the Mississippi Valley, thus splitting the Confederacy.

It was inevitable that the Mississippi should become the chief arena of conflict in the West. To open the Mississippi and separate the states of the Confederacy lying west of it became the chief aim of all the federal armies beyond the Alleghenies. The first effective step was taken when Grant forced the Confederates from all their strong positions on the Kentucky side of the Mississippi by his successful flank movement up the Tennessee and the Cumberland—

one of the most far-sighted strategical maneuvers executed during the war. The cleavage begun at Forts Henry and Donelson in February 1862 was completed at Vicksburg in July 1863.

Late in January 1862, Grant conceived the idea of breaking the eleven-mile Confederate line between Fort Henry and Donelson. He moved up the Tennessee River to Fort Henry with 17,000 men on transports escorted by seven gunboats. The garrison of Fort Henry, less than 3,000, retreated to Fort Donelson. Grant marched his army with difficulty across the watershed between the Tennessee and the Cumberland, and, on February 12, arrived before Fort Donelson with 15,000 men. This force was later increased to about 27,000. The Confederates holding the fort now numbered about 21,000 men commanded by General John B. Floyd with General Gideon J. Pillow and General Simon Bolivar Buckner as his chief subordinates. While awaiting the arrival of the gunboats which had to steam around from the Tennessee, Grant invested the place on the west and south. On February 13 a federal brigade attacked the batteries near the center of the line and was repulsed with heavy losses. The next day Commander Foote arrived with the gunboat flotilla and attacked the Confederate water batteries from the river. After a fight of two hours in which every federal vessel was more or less seriously damaged, Foote, himself wounded, retired. Grant, much disappointed, concluded that he would have to resort to a siege, though ill prepared to do so. The Confederate commanders, fearing to be trapped behind their own fortifications, determined to cut their way out and escape to Nashville. Early on February 15 they attacked the federal right

Dover Hotel, Scene of Brigadier General Simon B. Buckner's surrender of the Confederate Army at Fort Donelson to Brigadier General Ulysses S. Grant on February 16, 1862 (Fort Donelson National Park Service Photo)

flank and drove it back to the river. The road was now open to a Confederate retreat. This moment was the crisis of the battle. A good leader might have saved his army by immediate retreat or he might take advantage of the break in the federal line to throw his entire force in and boldly try to win a victory. Floyd did neither. In indecision, he permitted Pillow to order the victorious left wing to return to the trenches to cover a federal movement directed against another part of the line.

That night, a final council of war was held in the Confederate camp resulting in a decision to surrender. Floyd, who had been Secretary of War of the United States and was at this time under indictment in Washington, declared that personally he dare not surrender. For political reasons Pillow also did not care to fall into federal hands. They relinquished their commands and made good their escape in boats up the Cumberland. Floyd took with him the Victoria troops of his own brigade. One other

commander also made his escape but he took his whole command with him. That was Colonel Nathan Bedford Forrest, destined to win world fame as a daring leader of cavalry. Buckner, now in command, requested a truce. Grant then sent his famous ultimatum demanding "unconditional and immediate surrender." Buckner delivered to Grant between 12,000 and 15,000 officers and men as prisoners of war. The federal losses were about 5,000 killed and wounded and 450 missing.

The capture of Fort Donelson was Grant's first major victory in the Civil War and it greatly raised his prestige as a military leader. By strengthening the morale of the North it also has been rightly called one of the turning points of the war.

Today, you can visit the well-preserved fort, earthworks, rifle pits, and water batteries. Markers and tablets trace the course of the conflict. The National Cemetery contains 670 federal dead, 512 of them unknown, who were taken from their original graves on the battle-field and reburied here. All the facilities are administered by the National Park Service.

Fort Donelson is near the town of Dover, which is at the junction of State Routes 49 and 76, about eighty-seven miles from Nashville. Open daily from 8:00 A.M. to 4:30 P.M. No admission fee.

Fort Fisher

KURE BEACH, NORTH CAROLINA

FORT FISHER, named in honor of Colonel Charles F. Fisher, was the largest Civil War earthwork fortification in the Confederacy. On April 24, 1861, Confederate Point (now Federal Point) was taken over by the Confederates, and sand batteries mounting seventeen guns were constructed. Colonel William Lamb assumed command of the fort on July 4, 1862. Under his direction five hundred Negro laborers, assisted by the garrison, constructed a new and powerful Fort Fisher mounting forty-seven heavy guns. The heaviest naval bombardment of land fortifications up to that time took place there on December 24–25, 1864, and January 13–15, 1865.

Fort Fisher was important to the South because it kept Wilmington open to the outside world until the last few months of the Civil War. The swift blockade runners ran through the Union blockading squadron, providing the Confederacy with a vital supply of provisions, clothing, and munitions of war. The Federals realized the importance of closing this port, but were deterred from this action until late in the war because of the lack of a combined army-navy force large enough to capture and occupy the lower Cape Fear River area.

On the night of December 23, 1864, the Federals began their attack on Fort Fisher by exploding the powder ship *Louisiana* within three hundred yards of the fort. The 215 tons of powder did no damage. The federal fleet of fifty-six warships commanded by Rear Admiral David D. Porter bombarded the fort on December 24 and 25. On the afternoon of December 25 about 2,000 men from Major General Benjamin F. Butler's command of 6,500 troops were landed, but deciding that the fort was too strong to assault, the troops and the fleet withdrew.

On the night of January 12, 1865, the federal fleet reappeared numbering fifty-eight warships mounting 627 guns which shelled the fort continuously from the thirteenth to the fifteenth. The federal infantry, commanded by Major General Alfred H. Terry, had been increased to 8,000 troops. They were landed and moved across the peninsula to a point two miles north of Fort Fisher, and entrenched. Leaving 4,700 men in these entrenchments, facing Lieutenant General Braxton Bragg's 6,000 Confederates in the Sugar Loaf Lines, General Terry moved 3,300 men against Fort Fisher.

On the afternoon of January 15, the fort, with its garrison of 1,900 men, was attacked on the beach side by 400 marines and 1,600 sailors armed with pistols and cutlasses. After sustaining heavy losses they retreated in disorder. This attack served as a decoy enabling the infantry attacking on the river side to break into the fort. Bloody hand-to-hand combat finally forced 1,100 Confederates to surrender. Federal casualties numbered 1,500.

The Confederates evacuated the lower Cape Fear defenses after the fall of Fort Fisher and

Fort Fisher Visitor Center-Museum (Photo: State Department of Archives and History, Raleigh)

concentrated their guns and men at Fort Anderson at Brunswick on the west bank of the river. This was a last stand to protect Wilmington. Following a combined naval and land assault, Fort Anderson fell on February 19, and Wilmington, the capital of Confederate blockade running, was evacuated on February 21.

Gun emplacements, traverses, and other typical segments of the fort have been restored. The Visitors' Center-Museum houses exhibits pertaining to the Civil War and Fort Fisher.

Fort Fisher is on U.S. 421, south of Carolina Beach, New Hanover County. Travelers from Wilmington can pick up U.S. 421 directly and follow it south past Carolina Beach, Kure Beach, and into Fort Fisher. Open Tuesday to Saturday 9:00 A.M. to 5:00 P.M.; Sunday from 1:00 to 5:00 P.M. No admission fee.

"Dedication" ceremonies at Ft. Fisher (Photo: State Department of Archives and History, Raleigh)

Fort Frederica

ST. SIMONS ISLAND, GEORGIA

A N OLD BRITISH fortification dating from the early days of Georgia, Fort Frederica symbolizes Great Britain's desire to occupy the coastal lands—lands claimed by the Spaniards who were well entrenched at St. Augustine in Florida. Fort Frederica was built in 1736, enlarged and strengthened during 1739-43, and said to have been "the largest, most regular, and perhaps the most costly, of any fortification in North America, of British construction."

Georgia, the youngest of the thirteen British colonies in North America, was founded under the leadership of James Oglethorpe in 1733. Arriving at Savannah, February 12, 1733, Oglethorpe spent a year working with the mother city and outlying settlements. In January of the following year he made a trip down the inland waterway to select the site for the fort he planned to build for the protection of his infant colony. On the western shore of St. Simons Island he picked a high bluff where the Indians had cleared a thirty-to-forty-acre field. Here the river approached the bluff and made two right-angled turns—a strategic site for a fort. Oglethorpe named this site Frederica, in honor of Frederick, Prince of Wales, father of George III. He then returned to England to get the settlers who would build the town and fort.

Great care was used to select settlers. Frederica was planned in England as a typical English village. Frederica, with its outpost, St. Simons, was the most southern settlement made until then by the British in North America. Before the settlers left England they signed Articles of Agreement to perform the duties for which they had been selected. Magistrates, constables, doctor, midwife, minister, hatter, tailor, dyer, weaver, tanner, shoemaker, cordwainer, saddler, sawyer, woodcutter, pilot, accountant, baker, brewer, blacksmith, brazier, miller, millwright, and wheelwright. Accompanying this group as missionaries of the Church of England were John and Charles Wesley, who were later to become the founders of Methodism.

The first group of settlers consisted of forty families, numbering forty-four men and seventy-two women and children. Sailing on the *Symond* and the *London Merchant*, convoyed by the British sloop-of-war *Hawk*, they made a tempestuous crossing and anchored off Cockspur Island on February 5, 1736. They landed at Frederica in small boats on February 18. The next day they started to work on the earth fort, and a little over a month later its battery of guns commanded the river. Adjacent to the fort they laid out a town with eighty-four lots, sixty by ninety feet. Soon there were shops lining the streets and houses built of wood, brick, or tabby—a mortar made of sand, lime, and oyster shells.

With the founding of Frederica, the soldiers of the Independent Company stationed at Fort Frederick, near Port Royal, South Carolina, were ordered to St. Simons Island to protect the colony. They were stationed at Sea Point on the southern end of St. Simons where they built a fortification known as Delegal's Fort. The colonists also built Fort St. George on Fort George Island near the north bank of St. Johns River and Fort St. Andrews on the northwestern shore of Cumberland Island.

Oglethorpe returned to England to secure a regiment of British troops to man the fortifications already built and other forts he planned. He was commissioned "General and Commander in Chief of the Forces in South Caro-

Facing southwest: Two 12-pounder cannons now overlook the Frederica River (Photo: Peggy Dixon)

lina and Georgia."

Returning to Georgia in 1738 with a regiment of 650 British soldiers, the general built another larger fort at the south end of St. Simons Island. Still other fortifications built on this southern frontier for British colonies in North America included a fort at Darien, a lookout at Pike's Bluff on St. Simons, outposts at the present site of Brunswick and at Hermitage on Turtle River, Fort William on the southwestern shore of Cumberland Island, and even a blockhouse on Amelia Island in Spanish Florida. Fort Frederica was headquarters for all these fortifications and became the springboard for attack and base for defense against Spanish Florida.

The struggle for control of the Georgia-Florida-Caribbean area between Spain and Britain is known as the War of Jenkins' Ear. It was to merge into the Continental War of Austrian Succession (1740–48), and it is also known in American history as a phase of King George's War. In preparation for the conflict, Oglethorpe made a treaty with the Indians to ensure their aid. He strengthened Fort Frederica with tabby works and further enclosed the entire town of Frederica.

In November the Spaniards killed two of the Darien Highlanders who were stationed on Amelia Island. Oglethorpe invaded Florida and

Inside Fort Frederica—the remains of the King's Magazine and a 32-pounder cannon in the background; the foundation of the three-story South Storehouse in the foreground (Photo: Peggy Dixon)

captured Spanish outposts. He then made preparations for a grand invasion of Florida in the hopes of capturing the great fortress of Castillo de San Marcos at St. Augustine. They besieged the fortress for twenty-seven days until the approaching storm season forced them to retire after doing no more than destroying some Spanish outposts. In reprisal, Montiano, governor of Florida, invaded Georgia with fifty-one ships and 3,000 men. After several bloody and defeating attacks that got the Spaniards within a mile and a half of Fort Frederica itself, the Spanish forces returned to St. Augustine.

After the temporary peace brought about by the Treaty of Aix-la-Chapelle (1748) Ogle-thorpe's regiment was disbanded and Frederica practically abandoned. The great fire of 1758 destroyed most of the buildings, and after the ending of the French and Indian War in 1763, the soldiers were withdrawn and the cannon removed.

Fort Frederica is now a National Monument. Excavations have unearthed long-buried foundations of dwelling houses in Frederica, and storehouses, the King's magazine, the guardroom, and the blacksmith's shop within the fort area were excavated. Remains of the town gate, the moat, the bastion towers, and barracks building can be seen. A visitors' center tells the story of the old fort.

Fort Frederica is on St. Simons Island, twelve miles from Brunswick, Georgia. It can be reached by the Brunswick–St. Simons Highway (toll bridge) or by the inland waterway

Open daily from 8:00 A.M. to 5:00 P.M.; Summer from 8:00 A.M. to 6:00 P.M. No admission fee.

Fort Frederick

BIG POOL, MARYLAND

FORT FREDERICK, one of Maryland's early landmarks, is one of the best preserved pre-Revolutionary forts in America. For two hundred years its great rock walls have withstood the onslaughts of both time and enemies. It was built in 1756 by Governor Horatio Sharp during the French and Indian Wars to provide shelter to inhabitants of outlying settlements, following General Braddock's defeat at the Monongahela River. A chain of such colonial forts once stretched along the entire eastern side of the Alleghenies.

Fort Frederick occupies a dominant position on North Mountain, one hundred feet above the Potomac River, which flows a quarter of a mile to the south. Its rough stone walls, four feet thick and twenty feet high, are laid in a 240-foot square with bastions at each corner. The only entrance is through a heavy gate placed between receding walls and facing the river.

In colonial days there were barracks for a garrison of three hundred men. As many as seven hundred settlers at a time are said to have found refuge in the fort during Indian uprisings.

Fort Frederick was remote from the seaboard fighting in the Revolutionary War and was used as a prison camp for captured Hessian troops.

When peace came to the nation and the frontier in 1791, Maryland sold the fort and the 150 acres surrounding it at public auction. The land reverted to farm use.

When the Civil War broke out in 1861, territory along the Potomac was again a battlefield and Fort Frederick was garrisoned by part of the 1st Maryland Regiment. A cannon was mounted in the south wall of the fort to protect the Baltimore and Ohio Railroad and the Chesapeake and Ohio Canal, and to command a road across the Potomac River in Virginia. Enemy attacks to dislodge the garrison were unsuccessful.

When peace came, the old fort area was once again abandoned to agricultural purposes, and this situation remained until 1912 when the legislature, mindful of reviving interest in Maryland's rich historical heritage, authorized the purchase of the site. Negotiations were not concluded until 1922, when it was deeded to the state of Maryland. By that time the old walls had decayed into piles of rubble, and the foundations of the barracks inside were buried under accumulations of earth.

After intensive research, the original plans of the fort were located and with the aid of the Civilian Conservation Corps labor the outside stone walls and part of the interior were restored. The preservation of the historic old bastion and grounds give a good visual reminder of colonial times. A museum of colonial design and native materials is on the ground to house exhibitions of Indian relics, early firearms, and household and farm equipment used by the pioneers who sought refuge in the fort.

Fort Frederick in Washington County is

only a few miles off U.S. 40. It is reached on State Highway 56 by turning south at either Clear Spring or Indian Springs. Open April 15 to November 1, daily from 10:00 A.M. to 9:00 P.M. No admission fee.

James Fort

JAMESTOWN, VIRGINIA

On MAY 13, 1607, three small, storm-wracked ships, four months out of England by way of the West Indies, were moored to the shore of a peninsula on the James River about forty-five miles up from Chesapeake Bay. They carried an expedition of 105 men whose purpose was to settle in the New World. The leaders decided to make this the site of James Fort, or Jamestown as it came to be called, their first home in the wilderness and the English toehold on the North American continent. Until the Pilgrims landed at Plymouth in 1620, it was England's only settlement along the Atlantic seaboard. Within a century it grew from a crude palisade fort into a busy community as the capital of the colony of Virginia.

The colonists of 1607 first touched the American mainland on April 26 at Cape Henry, Virginia, and then they sailed up the James River to establish their settlement. The land looked good but it proved deceptive. Much suffering, starvation, and death awaited them. The first few years, when more settlers arrived, were a continuing struggle to overcome hunger, sickness, the ever-present wilderness, and the Indians against whom the fort was built.

Jamestown reached its lowest ebb in the winter of 1609–10, the "Starving Time", when 440 of the 500 inhabitants died. Yet the belief that the settlement would succeed spurred the survivors on. The fort was maintained, the crops planted, and the hard lessons of frontier living were learned.

Stronger, more orderly government came af-ter 1610, and by 1614 "James Citie," as the town was often called, had "two faire rowes" of houses and a street. The settlers ventured outside the fort proper, and built farms that occupied all of the high grounds of the peninsula.

Jamestown never became a town of appreciable size, but it served for nearly a century (1607–99) as the principal town and seat of government of Virginia. When the seat of government was removed to Williamsburg in 1700, Jamestown declined sharply and was eventually abandoned. By the time of the American Revolution the area became farmland; subsequently it was reclaimed by the wilderness.

Early Jamestown has been reconstructed and it is part of Jamestown Festival Park, a year-round exhibit maintained by the Commonwealth of Virginia. The three-cornered James Fort which the settlers built has been reconstructed; so have full-scale replicas of the three ships that brought the settlers to Jamestown: *Susan Constant, Godspeed*, and *Discovery*. There is also the tower of Jamestown's Church of 1639, which once served also for protection against Indian attacks.

Three of America's most historic sites are linked with Jamestown Festival Park by the Colonial Parkway as it stretches twenty-four miles from the James River to the York: Jamestown, Williamsburg, and Yorktown. The park is only a mile from Jamestown and adjoins the Colonial Parkway and State Highway 31. Williamsburg is six miles away. You can get to the Park from Richmond or Newport

A view of James Fort of 1607, as reconstructed at Jamestown Festival Park, Jamestown, Virginia

Reconstructed Jamestown glasshouse of 1608 (Photo: National Park Service, Yorktown, Virginia)

News to Williamsburg via U.S. 60, Interstate 64, or State Highway 168, thence to the park via State Highway 31 or Colonial Parkway. It is open daily except Christmas and New Year's Day from 9:00 A.M. to 5:00 P.M. Adults $1.00. Children 50¢.

Fort Jefferson

GARDEN KEY, FLORIDA

LIKE A STRAND OF BEADS hanging from the tip of Florida, reef islands trail westward into the Gulf of Mexico. At the end, almost seventy miles west of Key West, is the cluster of coral keys called Dry Tortugas. In 1513 the Spanish discoverer Ponce de León named them Las Tortugas (The Turtles) because of "the great amount of turtles which there do breed." The later name, Dry Tortugas, warns the mariner that there is no fresh water here.

Past Tortugas the treasure-laden ships of Spain sailed, braving shipwrecks and pirates. Not until Florida became part of the United States in 1821 were the pirates finally driven out. For additional insurance of growing commerce in the Gulf, a lighthouse was built at Tortugas, on Garden Key, in 1825. Thirty-one years later the present 150-foot light was erected on Loggerhead Key.

In the words of the military in 1830 any country that controlled Tortugas could control navigation of the Gulf. Commerce from the growing Mississippi Valley sailed the Gulf to reach the Atlantic. Enemy seizure of Tortugas would cut off this vital traffic, and naval tactics from this strategic base could be effective against even a superior force. With these thoughts in mind, the United States, during the first half of the 1800s, began a chain of seacoast defenses from Maine to Texas, protecting all of our coastlines. The largest link was Fort Jefferson, half a mile in perimeter and covering most of sixteen-acre Garden Key. From foundation to crown its eight-foot-thick walls stand fifty feet high. It had three gun tiers designed for 450 guns, and a garrison of fifteen hundred men.

The fort was started in 1846, and, although work went on for almost thirty years, it was never finished. The U.S. Engineer Corps planned and supervised the building. Artisans imported from the North and slaves from Key West made up most of the labor gang. After 1861 the slaves were partly replaced by military prisoners, but slave labor did not end until Lincoln freed the slaves in 1863.

To prevent Florida's seizure of the half-completed, unarmed defense, federal troops hurriedly occupied Fort Jefferson (January 19, 1861), but aside from a few warning shots at Confederate privateers, there was no action. The average garrison numbered five hundred men, and building quarters for them accounted for most of the wartime construction. Little important work was done after 1866, for the new rifled cannon had already made the fort obsolete. Further, the engineers found that the foundation rested, not on a solid coral reef, but upon sand and coral boulders washed up by the sea. The huge structure settled, and the walls began to crack.

For almost ten years after the Civil War, Fort Jefferson remained a prison. Among the prisoners sent there in 1865 were the "Lincoln Conspirators"—Michael O'Loughlin, Samuel Arnold, Edward Spangler and Dr. Samuel A. Mudd. Normally Tortugas was a healthy post, but in 1867 yellow fever came. From August 18 to November 14 the epidemic raged, striking 270 of the 300 men at the fort. Among the first of the thirty-eight fatalities was the post surgeon, Major Joseph Sim Smith. Dr. Mudd, together with Dr. Daniel Whitehurst, from Key West, worked day and night to fight the epidemic. Two years later, Dr. Mudd was pardoned.

Because of hurricane damage and another fever outbreak, Fort Jefferson was abandoned in 1874. During the 1880s, however, the United States began a naval building program and the Navy looked at the Tortugas as a possible naval base. It was from Tortugas Harbor that the battleship *Maine* weighed anchor for Cuba, where she was blown up in Havana Harbor on February 15, 1898. Soon the Navy began a coaling station outside the fort walls. One of the first naval wireless stations was built at the fort in the early 1900s. During World War I, Tortugas was equipped for a seaplane base. But as the military moved out again, fire,

storms, and salvagers took their toll, leaving the "Gibraltar of the Gulf" the vast ruin it is today.

A footnote to the history and location of Fort Jefferson is that one of our great national wildlife spectacles occurs each year between May and September, when the sooty terns assemble on Bush Key for their nesting season. The terns come from the Caribbean Sea and west-central Atlantic Ocean and land by the thousands on Bush Key. The great man-o'-war, or frigate, bird congregates there during the tern season. It has a wingspread of about seven feet.

Getting to Fort Jefferson has to be by seaplane or boat. The area available for approaches is limited to within one nautical mile of the fort on Garden Key. Because of a summer tern colony, seaplanes are not permitted within three hundred yards of Bush Key. This is an isolated wilderness. National Park Service guides are on duty at Fort Jefferson to guide you, but you must provide your own food and shelter—no supplies are available.

Fort Jesup

MARY, LOUISIANA

FROM ITS FOUNDING in 1882, until its deactivation in 1845, Fort Jesup was the most southwesterly military outpost of the United States. Because of a dispute over the Texas–United States boundary, in 1806, Spain and the United States designated as a neutral strip an area thirty to forty miles wide extending eastward from the Sabine River and embracing most of the present western tier of parishes in Louisiana. Under the Adams-Onis Treaty of 1819, the United States acquired this strip, which had become a haven for outlaws and marauders who molested settlers emigrating to Texas, and moved swiftly to occupy and police it. Pending ratification of the treaty,

which occurred in 1821 the United States Government in 1820 built Fort Selden on the Bayou Pierre near its junction with the Red River just outside the strip on its eastern edge, and the following year made plans to set up another post nearer the Sabine.

In 1822 the Army abandoned Fort Selden. Lieutenant Colonel Zachary Taylor occupied the watershed between the Sabine and Red rivers and moved to a point twenty-five miles south-southwest of Fort Selden, where his troops built a group of log cabins. Within a few months Cantonment Jesup, as the post was first called, was the largest garrison in Louisiana, consisting of four companies of the 7th

Infantry under Lieutenant Colonel James B. Many. In 1827–28 the troops helped construct a military road 262 miles northwest to Cantonment Towson, in Arkansas Territory; it linked the two most southwestern outposts of the U.S. Army. In 1831–33 General Henry Leavenworth assumed command of the cantonment and garrisoned it with six companies of the 3rd Infantry. In 1834 Colonel Many again assumed command. In 1833 the government, recognizing the enlargement and expansion of the cantonment, redesignated it as the Post of Fort Jesup and created the 16,000-acre Fort Jesup Military Reservation.

After the Texas Revolution began, in 1835, reinforcements arrived at Fort Jesup. Major General Edmund P. Gaines assembled thirteen infantry companies at the fort and marched to the Sabine, where he founded a temporary post, Camp Sabine. From there he occupied Nacogdoches and remained until the independence of Texas was assured. In 1845, just after President Tyler ordered General Zachary Taylor, commander of Fort Jesup, to move an army into Texas in anticipation of a war with Mexico, the Army deactivated the fort. It is a Registered National Historic Landmark relating primarily to political and military affairs, 1830–60.

In 1957 the state of Louisiana created Fort Jesup State Monument. It consists of about twenty-two acres. The only original building is one of the log kitchens, which has been repaired, reroofed, and refurnished with period reproduction and authentic kitchenware. An officers' quarters, reconstructed for use as a visitors' center, contains exhibits that tell the story of the fort.

Fort Jesup is located in Sabine Parish on Louisiana 6, about seven miles northeast of Mary. Open Tuesday to Saturday from 8:30 A.M. to 4:30 P.M.; Sunday from 1:00 to 5:00 P.M.; closed on Monday. Adults 50¢. Children under 12 free.

Fort Loudoun

VONORE, TENNESSEE

THE ENGLISH FLAG still flies over Fort Loudoun, built by the colony of South Carolina in 1756. This citadel of seven hundred acres, called the King's Land, a tract ceded by the Cherokee to the English, now points up an important and colorful landmark.

Eastward from the pinnacle of Fort Loudoun are the ridges of the Great Smokies. There the lifeline for food and supplies wound through the forests and chasms of rock toward Fort Price George in South Carolina. The English soldier standing guard at the fort was the bulwark against French encroachment from the Mississippi Valley. There was bloody rivalry between England and France and their colonies for the Indian trade in the Mississippi basin and beyond the Appalachians. Trade and military posts lined the paths from the Atlantic seaboard, across the mountain ranges, along westbound rivers to the Indian settlements and hunting country. Fort Loudoun was one of those protective outposts.

Nine Cherokee towns bordered the river's south bank. Chota, the tribal capital, and the ancient town, Tenassee, from which the state and two rivers take their names, lie upstream. Tuskegee, the birthplace of Sequoyah, is within eyeview of the fort. Sequoyah formed a set of characters to express the sense and sound of the Cherokee language.

Fort Loudon (Photo: Delmont Wilson)

Fort Loudon (Photo: Delmont Wilson)

The French were constantly conspiring to turn the Cherokee against the English. In 1759 they persuaded the Cherokee to cut the South Carolina supply line. Cherokee relationship with the English at this point was very strained. The English were insisting on the Cherokee surrendering Indians who had killed white settlers on the frontier, and although these raids were carried on by only a small group of Indians, the Cherokee refused to turn their warriors over to the English for imprisonment or worse. The English in turn established an ammunition embargo and threatened even more drastic measures. The French took advantage of this precarious situation, and the Cherokee cut the supply line to the fort. The soldiers and their families starved for months before submitting to surrender. Captain John Stuart and two junior officers were received at Chota Council House to arrange terms. Speaking for the Cherokee was "Old Hop," their principal chief, and war chief Oconostota. The besieged garrison were assured that they would be given safe conduct to Fort Prince George. The good faith of this promise was broken by French partisans among the Cherokee. They attacked the retreating band from Fort Loudoun at Bell Town, fifteen miles southeast of the fort. Twenty-six were killed and the other soldiers and families were seized as prisoners. Many of these were ransomed by the colonies of South Carolina and Virginia. The Cherokee occupied the fort after the surrender.

Today, Fort Loudoun is administered by the Fort Loudoun Association. It is enclosed by a star-shaped stockade. Massive log gates bar the entrance from the river. Replicas of the cannon of 1776 are mounted on wooden carriages in the fort. Original foundation stones of fireplaces in soldier's barracks, guardhouse, commandant's quarters, mark the building sites.

Fort Loudoun is located off Highway 411 between Maryville and Madisonville. It is open daily and on holidays from 9:00 A.M. to 5:00 P.M. Adults 75¢. Children 50¢. Children under ten free.

Fort McAllister

RICHMOND HILL, GEORGIA

STANDING ON THE HIGH land above the south bank of the Great Ogeechee River is Fort McAllister, an outstanding example of the earthwork fortifications of the Confederacy.

The significance of the fort is twofold. Its successful resistance to the attacks of the monitorlike vessels of the Union Navy in 1863 demonstrated that earthern fortifications could stand up against the most powerful naval ordinance employed up to that time. The fort played an important role in General W. T. Sherman's siege of Savannah in 1864, following his march through Georgia to the sea. Its capture on December 13, 1864, enabled the Union Army to communicate with the Navy and obtain much-needed supplies.

The fort was named in honor of the McAllister family, who owned a plantation nearby. Commenced in the early summer of 1861, its battery on Genesis Point was the southernmost in the elaborate series of Confederate fortifications that guarded Savannah from attack by the sea. It protected the vital trestle of the Atlantic and Gulf Railroad upstream as well as the rice plantations along the Ogeechee.

During the early days of the war, General Robert E. Lee, as commander of the military department of South Carolina, Georgia, and

Fort McAllister (Photo: Microfilm Division, Department of State, Georgia)

East Florida, inspected the battery at Genesis Point and made several recommendations. Credit for the ultimate design goes largely to Captain John McCrady, a Confederate engineer. He took advantage of lessons learned by the Confederacy at Fort Pulaski at the mouth of the Savannah. Deemed impregnable, its brick walls were easily breached by rifled cannon in April 1862, in a siege which demonstrated the obsolescence of masonry fortifications in modern war. Fort McAllister, however, was a massive earthwork, its gun emplacements being separated by large traverses, several of which contained service magazines. A Union naval officer described the fort as "a truly formidable work, so crammed with bomb proofs and traverses as to look as if the spaces were carved out of solid earth."

The routine of garrison life at Fort McAllister was broken in June 1862 after the arrival of the Confederate blockade runner *Nashville.* Thwarted in an attempt to run the blockade into Charleston, the swift side-wheeler eluded her pursuer after a long chase and slipped into the Ogeechee River. The bottling up or preferably the destruction of the *Nashville* was the first order of the day of the Union naval authorities. In order to get at the ship it was necessary to first silence Fort McAllister. Attacks were made by Union gunboats on July 29 and again on November 19, 1862. The shells dug large craters but did no real damage to the earthen fort. On January 27, 1863, a stronger effort was made to reduce the Confederate battery. Early that morning an oddly shaped warship steamed up the Ogeechee and anchored 150 yards below the barrier of piling which obstructed the channel. She was the U.S.S. *Montauk,* the second of the monitor class of vessel built by the North. Her revolving turret housed an eleven-inch and a fifteen-inch Dahlgren, the latter being the largest gun ever mounted on a warship. The ironclad was under the command of John L. Worden, the officer who commanded the U.S.S. *Monitor* in her encounter with the C.S.S. *Virginia (Merrimack)* in Hampton Roads. Bradley Sillick Osborn, a northern war correspondent on the *Montauk,* wrote, "The mooted question to be settled, are ironclads equal to forts?"

For nearly five hours the *Montauk* threw at the Confederate battery the largest shells ever fired up to that time by a naval vessel against a shore work. They plowed huge holes in the parapets but caused no consequential damage or casualties. "We failed to estimate the power and durability of that fort," Osborn ruefully concluded.

The *Montauk* sustained only minor damage. The fifteen direct hits she suffered made only slight indentations in her armor. On February 1, 1863, she made another attempt to destroy the Confederate battery. The Confederate commandant of the garrison, Major John B. Gallie, was killed and the *Montauk* was struck forty-eight times, again with little damage.

The *Nashville,* unable to escape to sea, had been converted into a raider and renamed the *Rattlesnake.* On February 27 she came down the river in an effort to get out, but the blockading vessels forced her back. In rounding Seven Mile Bend, upstream from the fort, she ran aground on a mud bank. The following morning the *Montauk,* taking advantage of the situation, ascended the Ogeechee and anchored twelve hundred yards from the hapless vessel. A three-way engagement resulted, the *Montauk* firing upon the stranded steamer and the guns of Fort McAllister firing on the *Montauk* and the two gunboats that accompanied her firing on the fort. In a short time the confederate gunboat was in flames and at 9:35 her magazine blew up, rattling the windows of Savannah twelve miles away.

A little while later there was another explosion. The *Montauk* in returning downstream struck a mine in the channel and had to be beached for emergency repairs. She did not join in the engagement that took place on March 3, 1863, the heaviest in which the Confederate fort was involved. Early that day the ironclads *Passaic, Nahant,* and *Patapsco,* together with three wooden gunboats and three mortar schooners, moved upstream. The squadron was commanded by Captain Percival Drayton, a former Charlestonian who had remained true to the Union. The object was to capture Fort McAllister. After a seven-hour bombardment Drayton reported that no damage had been done which "a good night's work would not repair." The only fatality at the fort was "Tom Cat," the garrison mascot. The Union ironclads retired, not to return. Fort McAllister had afforded a good testing ground for the monitors in preparation for the great attacks on the Confederate defenses in Charleston Harbor a few weeks later.

Fort McAllister was not attacked during the next twenty-one months. Late in 1864 it came back in the spotlight with the arrival of General Sherman's Army of 60,000 men before Savannah after his march from Atlanta. Fort McAllister was an important objective to the Union forces in the siege of Savannah. So long as it dominated the Ogeechee River, Sherman's troops were unable to obtain supplies from the Union vessels waiting off shore.

On the morning of December 13, 1864, the 2nd Division, 15th Corps of the Union Army of the Tennessee, marched down Bryan Neck to take Fort McAllister from the rear. Due to the difficult terrain, it took several hours to deploy the nine regiments that General W. H. Hazen used in the assault launched late in the afternoon. It was witnessed by General Sherman from the top of Cheve's rice mill on the other side of the river. He later wrote, "It was the handsomest thing I have seen in this war."

The assault lasted fifteen minutes and McAllister fell. Hand-to-hand fighting took place within the fort. The garrison of 230 men, commanded by Major George W. Anderson, was overpowered; it did not surrender. The Confederate losses were 16 killed and 54 wounded. The Union loss was 24 killed and 110 wounded. Many of the casualities were caused by the mines planted by the Confederates on the western approaches to the fort.

The fall of Fort McAllister marked the end of Sherman's "March to the Sea." Communication was shortly opened, via the Ogeechee, between the Union Army and the fleet. The loss of the Confederate fort rendered the further defense of Savannah useless. Seven days later the city was evacuated by General William J. Hardee, CSA. Concerning Fort McAllister, Lieutenant Colonel Charles C. Jones, Jr., wrote in his history of the siege of Savannah, 1864: "The heroic memories which it has bequeathed and the noble part it sustained in the Confederate struggle for independence will not be forgotten in the lapse of years."

During the late 1930s Henry Ford who owned the site of the fortification undertook extensive restoration. In 1958 the International Paper Company, which purchased the property from Mr. Ford's estate, conveyed the historic site to the state of Georgia by deed of gift.

Under the supervision of the Georgia Historical Commission, the earthworks and bombproofs have been restored to a condition closely approximating that of 1863-64. The museum, which contains many mementos of Fort McAllister and of the *Nashville,* was completed in 1963, one hundred years after the great bombardments by the Union ironclads.

Fort McAllister is located ten miles East of Richmond Hill, a suburb of Savannah. It is open from 9:00 A.M. to 5:00 P.M. daily. No admission fee.

Fort McHenry

BALTIMORE, MARYLAND

THE PRESENT SITE of Fort McHenry was recognized early in the Revolutionary War as a strategic location for military defenses to protect the water approaches to Baltimore. Fort Whetstone, a temporary fortification with exterior batteries, was constructed here in 1776 and its presence deterred British cruisers operating in Chesapeake Bay from molesting the city.

In the 1790s when war with either England or France seemed likely, it was decided that Baltimore was sufficiently important to merit a more permanent defense. The federal government and the citizens of Baltimore both contributed funds for this purpose. The outer batteries were rebuilt and strengthened and a new fort was constructed called Fort McHenry in honor of James McHenry of Baltimore, sometimes secretary to George Washington during the Revolution and U.S. Secretary of War from 1796 to 1800. The new star-shaped fort was replete with bastions, batteries, magazines, and barracks.

Until the War of 1812, life at Fort McHenry was routine and uneventful. In 1814 British troops, fresh from the capture and burning of Washington, appeared at the mouth of the Patapsco River. A joint land and naval attack on Baltimore was planned and on September 12 a landing was made at North Point. Encountering only moderate resistance, the British forces advanced to within two miles of the city, where they awaited the arrival of the fleet before attempting to storm Baltimore's defenses.

At dawn on September 13, a British fleet of sixteen warships anchored about two miles below Fort McHenry and commenced a heavy bombardment of this key defense work. During the next twenty-five hours, between 1,500 and 1,800 bombs, rockets, and shells were fired by the British but they only inflicted moderate damage to the fort. Casualties were also low—four men killed and twenty-four wounded. Convinced that Fort McHenry could not be taken, the British canceled their attack on Baltimore and withdrew their forces. The siege was over—the city was saved.

Fort McHenry never again came under enemy fire, although it continued to function as an active military post for the next hundred years. During the Civil War the fort was used by the federal government as a prison camp for captured Confederate soldiers. From 1917 until 1923, a U.S. Army General Hospital was located here to serve the returning veterans of World War I. In 1925 Congress, recognizing the historical significance of Fort McHenry, set it aside as a National Monument and a Historic Shrine.

Fort McHenry will always be associated in the minds of Americans with our national anthem. "The Star-Spangled Banner" is actually an account of the emotions felt by Francis Scott Key as he witnessed the British attack on the fort. Key came to Baltimore to secure the release of a friend seized by the British. He remained on the deck of an English cartel ship throughout the bombardment anxiously watching the fort and its flag. When at dawn of September 14 he once again saw the flag waving over the ramparts—even as the British fleet prepared to leave—Key began writing "The Star-Spangled Banner" to express what he felt. In 1931 Congress declared his song to be the national anthem.

Today, the visitor can see the raised mound that is the remaining part of the dry moat that originally encircled the fort and protected many

Fort McHenry

of its defenders during the twenty-five-hour bombardment. Along the trail to the fort, leading from the visitors' center where exhibits and a film depict the history of the fort and the writing of "The Star-Spangled Banner," there are two markers identifying the site of a tavern of the early 1800s and the historic road—now reconstructed—that in 1814 led into Baltimore. You enter the fort through an arched sally port constructed after the battle, with underground rooms on each side. These rooms were originally bombproofs housing gunpowder, but during the Civil War, Confederate prisoners were held here. The parade ground contains cannon believed to have been used in the defense of Baltimore. One of the cannon bears the monogram of King George III of England. At the front of the parade ground is the flagpole site from which spot the forty-two-by-thirty foot battle flag flew during the bombardment and inspired Francis Scott Key to write "The Star-Spangled Banner."

The guardhouses on either side of the entrance date from 1835, and the cells where Civil War prisoners were held, from 1857. In a semicircle around the parade ground are the several buildings that served as living quarters for the soldiers of the fort. These buildings, like the other quarters at the time of the bombard-

ment, were one and a half stories high with gabled roofs and dormer windows and porchless. Each contains exhibits that help interpret the fort. Also on display is the E. Berkley Bowie collection of firearms, spanning the period from the mid-1700s to World War I.

During the bombardment, the powder magazine was struck by a 186-pound British bomb which did little damage since it failed to explode. The magazine was later rebuilt and enlarged to its present size. The restored post headquarters was once the quarters of the commanding officer. The end room, today interpreted as the officers' kitchen and dining room, was in 1814 a separate building that served as a guardhouse. On the outer grounds there are Civil War batteries which replaced the 1814 water battery of thirty-six cannon located between the ravelin and the waterline.

Fort McHenry National Monument and Historic Shrine is three miles from the center of Baltimore and is readily accessible over East Fort Avenue, which intersects Maryland 2. The fort is open seven days a week from 9:00 A.M. to 5:30 P.M.; from mid-June through Labor Day, the visiting hours are from 8:00 A.M. to 8:30 P.M. All buildings close one-half hour earlier than the normal closing times. No admission fee.

Fort Macon

ATLANTIC BEACH, NORTH CAROLINA

DURING THE DAYS of the small sailing vessel, the region about Beaufort was highly vulnerable to attack from the sea. Old Topsail Inlet (now Beaufort Inlet) at the entrance to Bogue and Core Sounds is mentioned many times in pirate records. Edward Teach (Blackbeard) and other infamous characters passed through the inlet on their way to hiding places in the sounds where they rested their

crews and refitted their ships. The Spaniards during their height of sea power were a constant menace to the English colonists. In 1747 they captured the town of Beaufort, which they held for several days and plundered before being driven out.

North Carolina leaders early recognized the danger and sought to bar the entrance. The very year after the Spanish invasion, the legis-

Fort Macon was completed in 1834 and was garrisoned during four wars: the Civil War, the Spanish American War, and World Wars I and II. (Photo: Clay Nolen)

184

Fort Macon (Photo: Clay Nolen)

lature appropriated four thousand pounds for the building of four forts, one to be near Old Topsail Inlet, but this money remained unspent for several years despite intermittent raids on an impending French menace. Governor Arthur Dobbs, visiting the area in 1755 to select a site for a new fort, found that some work had been done on a fort on the mainland. He promptly decided that the fort should be built on the point of Bogue Island near Old Topsail Inlet. Accordingly a fascine fort, called Fort Dobbs, was begun there. Fort Dobbs was never completed, however, and the harbor remained defenseless until 1809. In this year, a small stone fort named Fort Hampton in honor of a North Carolina Revolutionary soldier was erected. It protected Beaufort during the War of 1812, but disappeared during the severe summer hurricane soon after the treaty of peace in 1815.

Fort Macon, the present fort, named for Nathaniel Macon, Speaker of the House of Representatives and United States senator from North Carolina, was begun in 1826 and was first garrisoned in December 1834. The garrison withdrew on February 2, 1836, as a result of congressional economizing but was regarrisoned during the years 1842–45, and again in 1848–49.

At the very start of the Civil War in April 1861, Fort Macon was seized by Confederate forces at the command of Governor John W. Ellis. It remained in Confederate hands until April 26, 1862, when federal forces under General John G. Parke captured it together with a Confederate garrison of approximately 450

commanded by Colonel Moses J. White. The fort was forced to surrender when federal guns located behind sand dunes, and about a mile down the beach, reinforced by those of four naval vessels standing off shore, heavily bombarded it. Thereafter the port of Morehead, with its railroad into the interior, was closed to the Confederacy.

During the Spanish-American War, Fort Macon was garrisoned but only two Parrot rifled cannon and two ten-inch mortars were used. The implacements for these guns can still be seen on the ocean side of the fort. Fort Macon was garrisoned by United States troops during World War II. On December 21, 1941, the 244th Division Coast Artillery occupied the fort because of "the importance of the harbor of Beaufort Inlet and the Navy installations under construction in that vicinity." The fort was returned to the state in 1946.

Fort Macon is a gem of early military architecture. The fort stands at the entrance to the Morehead City-Beaufort harbor. It has heavy gun emplacements, a moat, and a drawbridge common to fortifications of this period. It is now a part of the North Carolina State Park System. There is also a museum in the fort that visualizes the history, equipment, and life of the fort during various periods of its history.

Fort Macon is located in Carteret County on the eastern end of Bogue Banks. It is reached by turning south off of U.S. 70 in the western edge of Morehead City, crossing toll-free bridge over Bogue Sound, and turning east on the paved road. Open daily, hours not specified. Admission fee not specified.

Fort Matanzas

ST. AUGUSTINE, FLORIDA

THE BLOODY HISTORY of Fort Matanzas goes back to 1565 when between two and three hundred Frenchmen, all Huguenots, were put to the sword by the Spaniards. The Frenchmen had set out from Fort Caroline on the St. Johns River to attack the Spanish forces of Pedro Menéndez de Avilés. Menéndez, an able seaman and devout Catholic, had been sent by King Philip II to settle Florida. Jean Ribaut, a Huguenot, had been sent by Admiral Gaspard de Coligny, leader of the Huguenots, to establish bases within Spanish America in the name of the King of France. Fort Caroline was built on territory that the Spaniards considered their land. This trespass also threatened the route of merchantmen and treasure galleons returning to Spain from the Caribbean via the Gulf Stream. The French already had a long history of plundering Spanish ships.

On August 28, 1565, Ribaut arrived off the mouth of the St. Johns River. That same day Menéndez made his landfall at Cape Canaveral, 160 miles to the south. After two weeks of jockeying for position each side knew that confrontation must come. Ribaut took the initiative but a hurricane wrecked and scattered his ships far down the coast. Taking this chance, Menéndez marched overland through heavy rains and captured Ribaut's base. The French surrendered and the Spanish slaughtered them. Matanzas, which means "slaughters," had received its name.

For more than a century and a half the Spaniards controlled Florida. In 1773 General James Oglethorpe founded the English colony of Georgia on land claimed by Spain. Hostilities were inevitable, and the War of Jenkins's Ear between Spain and England gave him an excuse to attack St. Augustine. On June 13, 1740, Oglethorpe began the siege of St. Augustine by blockading the Matanzas River. Anticipating Oglethorpe's attack, Governor Manuel de Montiano had sent a courier to Havana asking for supplies, for they had enough for only three weeks.

On July 7 a courier reached St. Augustine and told Montiano that six supply ships were at Mosquito Inlet, sixty-eight miles farther down the coast. He also told them that the British had withdrawn the vessel blockading Matanzas Inlet, and the way appeared clear to provision the city. But simultaneously, an English deserter reported that Oglethorpe planned a night attack during the next six days of unusually high tides, for the high water was needed to cross Matanzas Bay and advance upriver.

Six days passed and no attack came, so Montiano sent five small vessels to Mosquito Inlet to fetch supplies. Just as the ships cleared Matanzas Inlet at four o'clock that afternoon, they met two British sloops that were taking soundings. The sloops opened fire and took up the chase. The fighting continued until twilight when the British sloops returned to their squadron. Their withdrawal gave the Spanish flotilla the opening they needed. They promptly entered Mantanzas Inlet, sailed up the river, and docked at St. Augustine that same night to the joyous relief of the inhabitants.

Fearing the approach of the hurricane season, the British fleet decided to sail for safer waters. Lacking naval support and knowing that the city was now well supplied, Oglethorpe raised the siege on July 20, 1740.

The British siege of 1740 convinced Governor Manuel de Montiano that he needed to have more than just a wooden tower at Matan-

zas Inlet. Had the British been able to seize that point, they would probably have been able to starve the city into surrender. Montiano therefore ordered engineer Pedro Ruiz de Olano to build a strong stone tower. Craftsmen came from St. Augustine and convicts and royal slaves did the heavy labor. Stone was quarried on Anastasia Island.

Construction was difficult, for long piles had to be driven deep into the mud to support the rising stonework. Repeatedly the British and their Indian allies tried to stop construction, but their efforts were in vain. By the end of 1742, work was complete. Next year the British attacked again, but heavy seas foiled their efforts. They withdrew, never to return.

Though British plans to acquire Matanzas by conquest always failed, they did gain all Florida by treaty in 1763. They, too, regarded Matanzas Inlet as the key to St. Augustine and usually kept seven soldiers and two cannon there, but no attacks came. Spain had planned to capture Matanzas and advance upriver to the Castillo de San Marcos during the American Revolution, but these plans never materialized.

With the passage of time the tower began to fall into disrepair. By 1821 the interior was already in ruins, and the gun platform's east wall and its foundation had cracked. Little interest was taken in the tower after the United States took control. Blockade runners used Matanzas Inlet during the Civil War, and for barely a decade the inlet was a port of entry. This activity, however, had little effect on the old tower, for soon the area was abandoned. In 1924 the fort was designated a National Monument.

Today, Fort Matanzas is part of a park of 298 acres on Rattlesnake Island, where the fort is located, and on Anastasia Island, where the visitors' center is. Landing docks for small craft are at both locations. The fort is accessible only by boat. A ferry crosses to Rattlesnake Island daily except Tuesday. State Road A1A from St. Augustine south reaches Fort Matanzas. The park is open daily from 8:30 A.M. to 5:30 P.M. except Christmas Day. No admission fee.

Fort Monroe

FORT MONROE, VIRGINIA

THE SITE of Fort Monroe was no stranger to fortifications. Fort Algernoume occupied the same ground from 1609 to 1667. In 1727 Fort George was built there and destroyed by a hurricane in 1749. During the War of 1812 the ineffectual system of coastal defense allowed the British to sack Hampton and sail up the Chesapeake Bay to capture Washington. The government then planned a new system of coastal defense which included Fort Monroe.

Construction of Fort Monroe began in 1819. Designed by General Simon Bernard, a French engineer, the fortification was placed to prevent the entrance of a hostile fleet into Hampton Roads and strong enough to withstand the attack of a combined naval and land force. Access to the inner fort could be reached by five entrances across the wet ditch, or moat, that completely encircled the structure. The buildings in existence at the time included quarters, barracks, and hospital. Because of its strength and position, the fort was sometimes called the "Gibraltar of Chesapeake Bay."

During the Civil War, Fort Monroe was one of the few forts in the South not captured by the

Aerial view of Fort Monroe

Confederates. It was a base for the Union Army and Navy and the scene of many dramatic events. General McClellan landed the Army of the Potomac at Fort Monroe in 1862 when he attempted to capture Richmond by advancing up the Virginia Peninsula. Abraham Lincoln spent May 6 to 11, 1862, at the fort, where he helped plan the operations against Norfolk.

General U.S. Grant was at Fort Monroe from April 1 to 3, 1864, to plan the campaign that finally won the war. The Army of the James, which played an important part in the Petersburg Campaign, was assembled at the Fort. Amphibious expeditions organized at Fort Monroe won strategic footholds along the Confederate coast from North Carolina to Louisiana, gradually closing down the major ports of the Confederacy.

As an engineer in the United States Army, Robert E. Lee assisted in the building of the fort and knew its massive strength—perhaps one of the most compelling reasons why the fort was never attacked during the Civil War. Lee, whose first son was born at Fort Monroe, served at the fort for three and a half years, from May 1831 to November 1834.

Jefferson Davis, the President of the Confederate States, became America's most famous political prisoner as the result of his imprisonment in a casemate (a chamber in the wall of a fort) at Fort Monroe. The cell, with its whitewashed stone walls and barred windows overlooking the waters of the moat surrounding the fort, was his jail for four and a half months. Lieutenant Colonel John J. Craven, chief medical officer at the fort in 1865, befriended the

captured Confederate President and gradually succeeded in relieving the harsh conditions of the casemate confinement. The cell can be visited today at the Casemate Museum located within the old fort's walls.

Another exciting event that took place near the fort was the four-hour battle between the two iron ships, *Monitor* and *Merrimack*. The Confederates, hoping to destroy the Union fleet in Hampton Roads and starve Fort Monroe into surrender, inflicted great damage with the ironclad *Merrimack* until the timely arrival of the *Monitor*. This first battle of the two innovative warships was a draw.

Of the originally intended 380 guns, several thirty-two-pounder cannon can be seen mounted in barbette. A tour of the fort takes the visitor to Quarters Number One. Built shortly after 1819, it is the oldest building at the fort. President Lincoln stayed here in May of 1862. Lee's Quarters is now a private residence. The Lighthouse, built in 1802, has been in continuous operation since that time. Lincoln Gun was cast in 1860 and named for the President in 1862. It is the first fifteen-inch Rodman gun and was used to bombard Confederate batteries on Sewell's Point. Outside of the fort are the water battery, redan, and redoubt.

The fort can be reached from Highway I-64. It is near Hampton Roads. Today it is still the home of the United States Army Training and Doctrine Command and a Registered National Landmark. It is open for visiting on weekdays from 8:00 A.M. to 5:00 P.M. and on weekends and holidays from 10:30 A.M. to 5:00 P.M. No admission fee.

Fort Moultrie

SULLIVAN'S ISLAND, SOUTH CAROLINA

THREE GENERATIONS before the Civil War—in another revolution—at the first of three forts built on this site, Colonel William Moultrie and about four hundred South Carolinians beat off a squadron of British warships on June 28, 1776. The Battle of Sullivan's Island, or Fort Moultrie as it came to be called, was one of the most decisive engagements of the American Revolution. It kept the South free of British control for the next three years and allowed southern men and supplies to strengthen the patriot cause in the North.

The present Fort Moultrie was built between 1807 and 1811. Its low fifteen-foot-high walls, covering one and a half acres, were built of sand faced with brick inside and out. Full

Fort Moultrie

armament was about forty guns. Three brick barracks built within the courtyard housed up to five hundred men, and a powder magazine held up to five hundred barrels of gunpowder. A furnace used to heat solid shot to red glow was also built in the courtyard. The barracks and furnace were destroyed in the Civil War, but the fort's original walls and powder magazine stand intact.

In the mid-1830s, while the United States Government was trying to relocate the Seminole Indians to open Florida to settlement, Osceola, a self-made leader of the Seminoles who opposed resettlement to the west, fought a two-year guerrilla war against settlers and the U. S. Army. Finally captured, he and two hundred other Indians were confined at Fort Moultrie to isolate them from Seminoles still fighting the war. Osceola died in Moultrie from malarial complications after only one month of confinement. Before his death, however, he had become a celebrity to Charlestonians and was granted considerable freedom within the fort.

New improvements in naval and coastal artillery were made after the Civil War, among them the development of breech-loading, rapid-fire guns. Battery Jasper was built in 1896 to hold these powerful new weapons. The huge concrete structure would also provide protec-tion against increasingly powerful naval armament. Although never tested in battle, such coastal batteries played a substantial role in safeguarding the United States from enemy attacks.

After World War II, when new weapons had completely transformed tactical and strategical concepts, forts like Moultrie became obsolete; and in 1947, after 171 years of service, Fort Moultrie was officially deactivated. The National Park Service took over its administration in January of 1961.

To commemorate the American Revolution Bicentennial, Fort Moultrie is undergoing extensive restoration. When the fort is restored as planned, it will tell the story of the changes that took place in fortification and armament over the years. Along the sea walls will be three batteries re-creating the look of the fort in three other periods of history, the Spanish-American War/World War I period (1898–1939), the post-Civil War period (1870s), and the Civil War period (1861–65).

Fort Moultrie is on west Middle Street on Sullivan's Island. From U. S. 17 take South Carolina 703 to Middle Street. The fort is open from 9:00 A.M. to 5:00 P.M. in winter and 9:00 A.M. to 6:00 P.M. in summer; closed Christmas Day. No admission fee.

Fort Osage

OLD SIBLEY, MISSOURI

Fort Osage may well rank with Jamestown and Plymouth Rock as a symbol of American beginnings. Upon the return of Lewis and Clark from the Pacific in 1806, the explorers were placed in charge of the newly acquired Louisiana Territory. Their problems were many. British traders encroached from the north; the Spanish were trying to ally with the Indians on the south and the rapidly growing fur trade and the protection of settlements from the Osage and other tribes added to their problems. Similar situations had been met since Washington's time by the operation of fortified trading houses on the Indian frontiers. So it was that Fort Osage came to be built by William Clark three hundred river miles from the

Blockhouse of Fort Osage (Photo: Missouri Division of Tourism)

nearest white habitation.

The building party consisted of a company of the 81st U. S. Infantry and a company of the 80th St. Charles Dragoons mounted militia. They arrived on the site September 4, 1808, and the post was christened Fort Osage on November 10, 1808. Its purpose was to establish friendly relations with the Indians by giving them a trading post run by the government, enforce the licensing of private traders, and serve notice upon the British and Spanish colonial authorities that the United States would resent encroachments upon its new territory. The fort was built of hewn white-oak logs on a high promontory overlooking the Missouri.

Fort Osage has unique historic interest, not only because it was the first United States Army post beyond the settlement along the Mississippi but also because of its associations with the earliest personalities and movements identified with the opening of the West. The logic of its location near the fork of the northwest and southwest routes around the Rockies and was proved by the later development of nearby Independence, Westport, and Kansas City, as termini of successive methods of continental transportation. Fort Osage was among the most successful of twenty-eight such posts operated one time or another under the United States system.

While supervising the building of the fort, William Clark required the Great and Little Osage to move their villages to the immediate vicinity and obtained the cession to the United States of all their land to the east and between the Missouri and Arkansas rivers. There were at one time some five thousand Indians about the fort. In April 1811 an expedition organized by John Jacob Astor left the fort on the first expedition, after Lewis and Clark, to reach the Pacific. They built a fur post at the mouth of the Columbia River. They were followed by the annual Missouri Fur Company party. The

War of 1812 caused the evacuation of Fort Osage during the years 1813–15, five of the northern U.S. forts having been destroyed by British and Indians.

Garrison life resumed in 1816. Factor Sibley, who was in charge of the fort's trading post, or "factory" as the system was known, was host to Daniel Boone, then in his eighty-fifth year. The Army's Yellowstone expedition of 1819 brought the first steamboats and removed the garrison to establish Fort Atkinson above present Omaha. In 1821 the first successful trading party to Santa Fe stopped at the fort, and in 1825 George C. Sibley was appointed one of the commissioners for federal survey of the Santa Fe Trail beginning at the gate of the fort. By the Osage treaty of 1825 the Indians ceded their remaining land in Missouri. Previously the trading house, or the United States factory system, was abandoned through opposition of the fur companies, and the usefulness of Fort Osage was ended.

Fort Osage has been restored to preserve the opening history of the Louisiana Purchase and to present an example of the United States factory system which played an important part in the development of Indian affairs of the nation. The visitor can now see Blockhouse No. 1, the largest of five that defended the fort and whose guns could command the river. Cannon and historical exhibits are on view. The factory itself has been restored and its equipment, furniture, and furnishings are on view. It is constructed of hewn white-oak logs from the Ozark Mountains and finished with hand-wrought hardware and window glass. A museum also illustrates the operation of the factory and its relation to the first two decades of the Louisiana Purchase.

Fort Osage is at Old Sibley, on the Missouri River, fourteen miles northeast of Independence. Turn north from U. S. 24 to 20E at Buckner and follow signs, or take the old Blue

Mills Road, 8N, leaving U. S. 24 at Salem Church, four miles east of Independence. Continue to the Buckner Road, 20E, and follow signs. It is open during "daylight hours." No admission fee.

Fort Pickens

PENSACOLA BEACH, FLORIDA

OLD BRICK FORT PICKENS, built on the western tip of Santa Rosa Island to defend the deep-water harbor of Pensacola against foreign attack, has been the scene of activities in every conflict in which this country has been engaged, from the Civil War to World War II.

Its major role was during the Civil War, when it remained one of the three southern forts (all in Florida) that the Confederates were unable to seize from the Federals.

Santa Rosa Island has been the location of many fortifications since the early 1700s. It was recognized by the Spanish, French, British, Confederates, and the United States Army and Navy as the key to the defense of Pensacola Harbor.

Pensacola was the capital of West Florida when the Florida territory was ceded to the United States in 1821. In anticipation of the selection of Pensacola as the principal United States naval depot of the Gulf of Mexico, the

The flag still flies over old Fort Pickens (Photo: Florida Park Service)

U. S. Government built four forts in the area. Fort Pickens, completed in 1834, was named in honor of Brigadier General Andrew Pickens of the South Carolina State Troops in the American Revolution.

At the onset of the Civil War, the fort had been unoccupied for several years. The day Florida passed her ordinance of secession, January 10, 1861, the fort became the federal headquarters. Confederates made several attempts to capture Fort Pickens, but all hostilities were stopped when they evacuated their holdings in the area in May of 1862. The Federals immediately took possession of the other deserted forts and the Navy Yard, and held Pensacola for the rest of the war.

As the war waned, Fort Pickens was used as a prison for military and political prisoners. In 1875 Congress again used Fort Pickens as a prison with the Apache Geronimo and some of his band being held there for two years after their capture in 1866.

Fort Pickens was an active coastal defense fortification during the Spanish-American War, and was also activated during the two world wars.

Today you can see evidence of the struggles for possession of the strategic area still to be found within the Fort Pickens State Park. There are three small ponds located between the park entrance and the gatehouse, where the Confederates landed and were defeated in the Battle of Santa Rosa Island. All of the battlefield is within the park. You can visit Battery Langdon, built in 1917 and reconstructed in 1942. There are other fortifications including British and Civil War batteries. Old British and American cannons are on display.

Fort Pickens is located seventeen miles from Pensacola on State Road 399 at Pensacola Beach. A standard fee is charged at all Florida State Parks and there are facilities for tent and trailer camping.

Fort Pulaski

SAVANNAH BEACH, GEORGIA

TWENTY-FIVE MILLION BRICKS, one million dollars, and eighteen years of toil went into the construction of Fort Pulaski. Construction started in 1829 by United States military engineers on a low, grassy island to guard the river approaches to Savannah. The latest design in fort construction, it was said at the time to be "as strong as the Rocky Mountains."

Why was this great fort built? In 1815 the United States stood on the threshold of a fabulous century of progress. The country had fought England to a draw in the War of 1812, and the memory of victories on land and sea was like heady wine. Yet some people also remembered the disasters.

President Madison was one of those who remembered how easily England had penetrated our unguarded shores, raided farms and hamlets along the Chesapeake, and left the nation's capital in smoking ruins. To prevent such humiliation in the future, he commissioned Simon Bernard, a famous French military engineer, to plan an invincible system of defense.

Fort Pulaski was one link in the chain of seacoast forts recommended by Bernard. The fort, named for Count Casimir Pulaski, a Polish soldier of fortune who lost his life in the unsuccessful siege of Savannah in 1779, stands

on Cockspur Island at the mouth of the Savannah River in Georgia. The fort was the pride of Savannah, and all important visitors were taken to see it. But when tested in battle during the Civil War, it failed to meet the expectations of its admirers. Subjected to bombardment by U. S. troops, the Confederate defenders surrendered in thirty hours.

The quick fall of Fort Pulaski surprised and shocked the world. There was no mystery in the fall of Fort Pulaski. It was the old story of a new and more powerful weapon overcoming an obsolete system of defense. The battle, brief and undistinguished though it was, proved the superiority of rifled cannons over masonry forts. In thirty hours, ten of these new weapons breached the seemingly impenetrable walls and proclaimed to the world that the day of brick forts had ended.

A brief recapitulation of the defeat of Fort Pulaski is in order. On January 3, 1861, when relations between North and South were strained to the breaking point, Georgia troops seized Fort Pulaski. South Carolina had already seceded from the Union, and it seemed inevitable that Georgia would follow.

At the time of this break Savannah was a city of about twenty thousand inhabitants and a rich seaport trading in cotton, naval stores, and timber. Its leaders were citizens of wealth and culture. Though they were divided on the strategy of seizing a federal fort, nevertheless they all joined in preparations for its defense after the seizure. Georgia seceded on January 19, 1861, and a few weeks later transferred Fort Pulaski to the Confederate States of America. By the end of April, eleven southern states had left the Union and were at war with the United States.

During the first year of the war the South was riding high, but before the end of the summer, President Lincoln ordered the Navy to blockade southern ports. As the blockade tightened it strangled the Confederate economy. On November 7, 1861, a combined army and navy expedition struck at Port Royal, South Carolina, about fifteen miles north of Fort Pulaski. The heavy guns of the Federal warships bombarded Fort Walker. Unaccustomed to battle, the Confederates retreated in disorder, and the Federals landed unopposed on Hilton Head Island. From this beachhead deep in southern territory, the Federals immediately established a base for operations against Fort Pulaski and the entire South Atlantic coast.

The defeat at Hilton Head convinced the Confederates that it would be impossible to hold a defensive line on the sea islands. On November 10 they abandoned Tybee Island at the mouth of the Savannah, burned the lighthouse, and destroyed their earthworks. This action proved to be a great error, for it gave the enemy the only site from which Fort Pulaski could be taken. The Federals moved quickly to take advantage of the break. They first cut the fort's communications with the mainland, then moved in force to Tybee Island to prepare for battle.

Engineer Captain Quincy Adams Gillmore, commander of the siege operations, believed that an overwhelming bombardment would force the Confederates to surrender the fort. He erected eleven batteries containing thirty-six guns and mortars along the northwest shore of Tybee Island. On April 10, 1862, he sent a courier to Fort Pulaski under a flag of truce with a formal demand to surrender.

When the Confederates refused, the Federals opened fire. But the Confederates were not particularly alarmed. Federal guns were a mile or more away, over twice the effective range for heavy ordinance of that day. What the fort garrison did not know was that the federal armament included ten new rifled guns destined to make military history. Soon projectiles from these guns began to bore through the bricks of Pulaski with devastating effect. By noon of the second day the bombardment had opened wide gaps in the walls, and explosive shells, passing through the gaps, threatened the main powder magazine. Confederate Colonel Charles H. Olmstead considered the situation hopeless and gave the order to surrender.

Gilmore was the hero of the day. His boldness in using the new weapons won him a victory and he was breveted a brigadier general. To the 1st Connecticut and 3rd Rhode Island went the honor of receiving the surrendered fort. The captured garrison numbering 385 officers and men were sent north to Governor's Island in New York. The 48th New York Volunteers were assigned the job of repairing the damaged fort. During the last days of the war Fort Pulaski was used as a military and political prison. A final salute of two hundred guns was fired from the fort's ramparts on April 29, 1865, to commemorate the end of the Civil War.

Fort Pulaski resembles a medieval castle. It is surrounded by a wide moat, with two drawbridges, and a rear fortification known as a demilune. After crossing the outer drawbridge,

a short walk through the demilune will bring you to the second drawbridge, and the sally port or only entrance into the main fortification. Numbered markers will take you to the sally port, the gorge (rear section of the fort), the barracks rooms, the north magazine, the northwest bastion (the part of the fort that extends out from the main wall, with embrasures and loopholes to permit lateral fire along the walls), the gun galleries, the water system, the terreplein (flat surface on top of the rampart), the prison, the breach, the southwest bastion, the headquarters, including the Surrender Room, actually the quarters of Colonel Olmstead, the Cistern Room, the moat, Damaged Wall (evidence of the devastating bombardment), and the cemetery.

Fort Pulaski, on Cockspur Island at the mouth of the Savannah River, is fifteen miles east of Savannah, Georgia. It may be reached from that city by way of U. S. 80 (Tybee Highway). The entrance is on McQueens Island at U. S. 80. Cockspur Island is connected by a short road and a concrete bridge across South Channel of the Savannah River. It is open daily from 8:30 A.M. to 5:30 P.M. Admission fee is one dollar per carload.

Fort Raleigh

MANTEO, NORTH CAROLINA

THE NORTH END of Roanoke Island, North Carolina, is the scene of Sir Walter Raleigh's ill-fated attempts to establish an English colony in America. It is our connecting link with the Court of Queen Elizabeth and with the golden age of the English Renaissance. In 1583 Sir Humphrey Gilbert, half brother of Sir Walter Raleigh, staked all he had in an attempt to found a colony in the northern part of North America. It was not successful and he

himself was drowned on the return voyage to England. Imbued with a desire to realize his brother's dream of an English empire in America, Raleigh sent Captains Philip Amadas and Arthur Barlowe to America in 1584 to select a site for a colony. They explored the North Carolina coast, including Roanoke Island, and returned with a favorable report on the island which they described as "a most pleasant and fertile ground." In honor of Queen Elizabeth,

the whole country was named Virginia.

Raleigh's first colony, consisting of 108 persons, departed from Plymouth, England, April 9, 1585, under the command of Raleigh's cousin Sir Richard Grenville. A settlement was made on the north end of Roanoke Island. Ralph Lane, a relative of the English royal family, was made governor when Grenville went back to England for supplies. Lane built Fort Raleigh, calling it simply "the new Fort in Virginia." Dwelling houses were built near the fort and, with the assistance of the Indians, crops were planted and fish traps made. The country was explored for a distance of about 80 miles to the south and 130 miles to the north. Thomas Hariot, the geographer, collected data for his *New Found Land of Virginia,* and John White, the artist, made watercolor drawings of the Indians, the animals, and plant life of the country.

But Grenville's supply was late in returning to Roanoke. Open war with the Indians ensued and food became scarce. When, on June 10, 1586, Sir Francis Drake, en route from the West Indies, anchored off the coast near Roanoke Island with a fleet of twenty-three ships, many of the settlers were dissatisfied with the hardships and wished to return home. Despite Drake's offer of assistance and supplies, he took the surviving members of the colony back to England.

Shortly afterward Richard Grenville arrived at Roanoke and found that the colony had gone. After searching for it elsewhere on the coast in vain, he left fifteen men on Roanoke Island with provisions for two years to hold the country for Queen Elizabeth, and returned to England.

Raleigh's second colony, consisting of 150 men, women, and children, arrived at Roanoke Island under the government of John White and twelve assistants in July 1587. They found only the bones of Grenville's men. The fort had been razed, but the houses were standing. The old houses were repaired and new cottages built. On August 13, pursuant to Sir Walter Raleigh's orders, the friendly Indian chief, Manteo, was baptized and created Lord of Roanoke. On the eighteenth, Eleanor, daughter of Governor White and wife of assistant Ananias Dare, gave birth to a daughter, who was christened Virginia, the first English child born in Virginia.

Governor White returned to England for supplies. He found England in imminent danger of invasion by Spain, and the Queen felt that the danger to England was so great that she could not spare any large ships for supplies to the colony. Two small pinnaces allowed to leave England never reached Roanoke.

When Governor White returned to Roanoke in August 1590, he found that the colony had disappeared. The houses had been taken down and the place of settlement enclosed with a high palisade, with curtains and flankers "very fort-like." One tree at the right side of the entrance to the palisade had the bark peeled off and the word "Croatoan" engraved on it. This was the distress signal agreed on in case of difficulties or enforced departure. White concluded that the colonists would be found on Croatoan Island (most of modern Ocracoke and part of Hatteras islands) south of Cape Hatteras, or among the Croatoan Indians farther inland. This mystery of the tragic "Lost Colony" has never been solved.

Fort Raleigh was designated a National Historic Site on April 5, 1941. With its nearly nineteen-acre area, parts of the settlement sites of 1585 and 1587 are included. Ralph Lane's "new Fort in Virginia," located within the site, was explored archaeologically in 1947–48 and restored in 1950. The village site, presumably close to the fort, has not yet been located. Excavated artifacts are displayed in the visitors' center.

Fort Raleigh National Historic Site is on State Route 345, three miles north of Manteo, North Carolina, ninety-two miles southeast of Norfolk, Virginia, and sixty-seven miles southeast of Elizabeth City, North Carolina. It may be visited daily except holidays in winter from 8:00 A.M. to 4:30 P.M. During the summer, the *Lost Colony*, a pageant-drama by Paul Green, is produced in the Waterside Theater, according to hours and dates fixed by the sponsoring Roanoke Island Historical Association. No admission fee.

Fort Sumter

SULLIVAN'S ISLAND, SOUTH CAROLINA

ON APRIL 12, 1861, a mortar shell fired from Fort Johnson in Charleston Harbor burst almost directly over Fort Sumter, and the American Civil War had begun. Fort Sumter is one of a series of coastal fortifications built by the United States after the War of 1812. The five-sided fort, erected on a shoal and bearing the name of the South Carolina Revolutionary War patriot, Thomas Sumter, was started in 1829, and essentially completed by 1860. The five-foot-thick walls, of brick, towered 48.4 feet above low water to command the main ship channel into Charleston Harbor. Four sides, 170 to 190 feet long, were designed for three tiers of guns; the gorge, designed for officers' quarters, supported guns only on the third tier. Enlisted men's barracks paralleled the parade side of the flank gun rooms. A sally port pierced the gorge and opened into a quay and a wharf. Full armament was about 135 guns, but by 1861 only sixty cannon had been placed.

On December 20, 1860, South Carolina became the first state to secede from the Union. By March 2 six other states seceded, a constitution was adopted, Jefferson Davis was elected President and inaugurated February 18. All the forts and navy yards in the seceding states were seized by the Confederate government, except Fort Sumter.

During the night of December 26–27, 1860, Major Robert Anderson moved his small federal garrison from Fort Moultrie, on the north side of the harbor, to Fort Sumter out in the harbor where they could not be reached by land. After completing the move, he reported to the War Department that he had about a four-month supply of provisions. Charlestonians immediately occupied Fort Moultrie and started to build batteries elsewhere about the harbor. General P. G. T. Beauregard was given command of the Charleston area for the Confederate forces.

Meanwhile, in January, President James Buchanan made an effort to reinforce and supply Fort Sumter. The *Star of the West* sailed from New York, unarmed, to the relief of Sumter. It was fired upon as it entered the harbor on January 9 by South Carolina shore batteries. The ships turned back for New York. When Lincoln took office he ordered supply ships to sail from New York to Sumter on April 6, and at the same time notified Governor Pickens of South Carolina that a relief ship could be expected. After debate, the Confederate Cabinet ordered General Beauregard to fire on Sumter if any federal ship tried to reach it.

On April 11 General Beauregard demanded that Sumter be evacuated. Major Anderson refused. An ultimatum was given Anderson at 3:20 A.M. on April 12 that the Confederate

Fort Sumter from the southeast (Photo: Fort Sumter National Monument)

206

batteries would open fire in one hour. At 4:30 A.M. a mortar at Fort Johnson fired a shell that was a signal for the bombardment. Within a few moments a gun directed at Fort Sumter was fired from the ironclad battery at Cummings Point, and by daybreak forty-three guns from Fort Johnson, Fort Moultrie, and batteries at Cummings Point and elsewhere were firing at the fort. Not until seven o'clock did the guns at Sumter reply. Ammunition was low, and by noon only six guns were kept in action. Only five men in the fort were wounded during the bombardment.

Throughout the night the firing continued. The next morning a hot shot from Moultrie started a fire at Sumter. This intensified Con-

federate fire. At about 2:00 P.M. an offer of truce was carried to Sumter. It was accepted and the fort was evacuated on April 14, after a defense of thirty-four hours. During that time the quarters were entirely burned, the main gates destroyed by fire, the gorge walls seriously damaged, and the magazines surrounded by flames. The Confederates had fired more than three thousand shells at Sumter.

For nearly two years, determined Confederates on Fort Sumter kept federal forces at bay. By 1863 the Federal Navy controlled all important Atlantic coastal ports except Charleston and Wilmington. An April 7, 1863, nine armored federal vessels under Rear Admiral S. F. DuPont steamed slowly into a long battleline and headed

Fort Sumter

for the fort. In the two-hour artillery duel that followed, the fort had its walls scarred and battered, but the attack failed. Five of the ironclads were disabled and the *Keokuk* sank the next morning. The summer of 1864 saw a renewed attempt to take Fort Sumter, this too failed. For twenty-two months Fort Sumter had withstood federal siege and bombardment. Thirty-five hundred tons of shells had been hurled at it. Fifty-two killed and 267 wounded were its casualties. Sherman's troops advancing north from Savannah, however, caused the Confederate troops to be withdrawn, and Sumter was evacuated on February 17, 1865.

Fort Sumter is a National Monument. It contains 2.4 acres and reflects many alterations and changes that have taken place in it since 1865, particularly the addition of Spanish-American War Battery Huger.

The fort stands on a shoal at the entrance to Charleston Harbor. Each day at 2:30 a Gray Line boat leaves for the fort from Murray Boulevard at the foot of King Street, Charleston. An additional trip is made daily at 10:00 A.M. March through Labor Day. There is no admission fee.

Fort Ward

ALEXANDRIA, VIRGINIA

FORT WARD was a minor bastion, typical of the many hurriedly constructed during the Civil War. When the Civil War began, the city of Washington was practically in front lines, completely unprotected against an attack by land. On the day that Virginia's secession became effective Union troops crossed the Potomac River, seized Alexandria and Arlington Heights, and then built three forts south of the river.

After the Confederate victory at Manassas (Bull Run), July 21, 1861, the North hastily began erecting defenses for the Capital on both sides of the river, encircling Alexandria, Washington, and Georgetown. Work on Fort Ward, named for Commander James Harmon Ward, the first Union naval officer killed in the Civil War, was begun on September 1, 1861. By the end of the year over forty forts had been built. Then all work ceased until the South won its second victory at Manassas in August 1862.

Attention was again focused on the defenses of Washington. They were enlarged, strengthened, and more fortifications erected until by late 1862, the city had become the most heavily defended location in the Western Hemisphere. It was completely surrounded by sixty-eight forts and batteries, mounting 905 guns, connected by over thirty miles of trenches and roads. Fort Ward, with its thirty-six guns and five bastions, was the fifth largest fort in this chain. After the war, these forts were either dismantled or abandoned.

Fort Ward was restored, with the help of Mathew Brady photographs, old records, and archaeology. To preserve the restoration the city of Alexandria purchased forty acres of woodland surrounding the fort and then carefully restored the northwest bastion to look as it did a hundred years ago. The guns mounted therein are exact replicas of guns used in Fort Ward during the Civil War. The officers' hut and the museum were constructed from photographs of wartime buildings located in Alexan-

Bastions and museum, Ford Ward

209

dria. The Museum contains one of the largest collections of Civil War items in the country.

Fort Ward is located on West Braddock Road at the top of the hill between King Street (State Route 7) and Seminary Road. It is just east of the Shirley Highway (U. S. Interstate 95). It is open Monday to Saturday from 9:00 A.M. to 5:00 P.M.; Sunday from noon to 5:00 P.M.; closed Thanksgiving and Christmas. No admission fee.

Fort Washington

OXON HILL, MARYLAND

In 1808, after repeated seizures of American seamen and the detention and search of the U. S. frigate *Chesapeake* by a British naval vessel, the government decided that a defense of its ports and harbors was necessary and that priority should be given to the new seat of government in the District of Columbia. Land on the Maryland shore of the Potomac opposite Mount Vernon was acquired from the Digges family of Warburton as a site for a fort. This location had been selected earlier by George Washington, in 1794, when the construction of a fort on the Potomac was under consideration.

Work on the first Fort Washington was completed December 1, 1809. It was described as an "inclosed work of Masonry" having a semi-elliptical face with circular flanks enclosed by a perpendicular wall suitable for defense by small arms. The height of the rampart was generally fourteen feet above the bottom of the ditch. The main fort was commanded by a "Tower of Masonry calculated to contain one company" and six cannon. This early fort stood for only five years. It was destroyed in August 1814 when the British successfully attacked the new Capital. The British offensive began on August 19 when their troops marched toward the Capital by passing Fort Washington. On August 24, 1814, they defeated the Americans in the Battle of Bladensburg and captured Washington. There, they burned the Capitol and the White House, and most of the other public buildings. British war vessels, moving up the Potomac to cooperate with their land forces, reached Fort Washington on August 27. Captain S. T. Dyson, who commanded the position, destroyed and abandoned the fort without offering resistance.

On September 8, 1814, only twelve days after the destruction of the first Fort Washington, Acting Secretary of War James Monroe requested the French engineer Major Pierre L'Enfant to reconstruct the destroyed fortification. He supervised the clearing away of the debris and demolition of the old fort, included 200,000 bricks and a large quantity of stone and lumber to begin work on a new water battery and wharf. The free hand given to L'Enfant during the emergency was now under scrutiny. Anxiety for the safety of the Capital relaxed after the Treaty of Ghent was received and the British ships in Chesapeake Bay sailed for Jamaica. After being questioned again and again about economy, L'Enfant resigned. On September 6, 1815, Lieutenant Colonel Walker K. Armistead took over the work.

The fort took nine years to be completed, and it has been little altered since 1824. It is an enclosed masonry fortification, entered by a drawbridge across the dry moat at the sally port. From above the main gateway you can see the entire 833-foot outline of the fort. Sixty feet below the main battery is the outer V-shaped water battery, begun by L'Enfant. Two half bastions overlook and command the river above and below the fort. Below the ramparts are the bombproof gun positions. From three levels guns could deliver a devastating fire against an enemy fleet on the Potomac. The front of the structure—built of solid stone and brick masonry—is about seven feet thick. On the parade ground are the officers' quarters and the soldiers' barracks. Flanking each of these structures is a magazine.

During the 1840s eighty-eight permanent gun platforms for barbette carriages were constructed, a new drawbridge built, and the powder magazines improved. The south wall was raised and strengthened by the addition of a bastioned outerwork. Physical evidence of all these improvements can still be seen. Most of the mechanism to raise the drawbridge remains in place.

During the Civil War, on January 5, 1861, the first order issued by the Secretary of the Navy for the defense of the national Capital assigned forty marines to protect Fort Washington, at that time the only fortification defending the city. During the war, troops from the 4th Artillery and other units manned the fort. Its importance decreased as attack by water became less probable.

Fort Washington was abandoned in 1872, and thirteen years later the obsolete muzzle-loading guns were removed. From 1896 to 1921 the reservation was headquarters for the Defense of the Potomac. During this period, eight concrete batteries were constructed near the old fort. These concrete batteries can still be seen, although the guns have been removed. In 1921 the fort became headquarters of the 12th Infantry, and in 1939 Fort Washington was transferred to the Department of the Interior. Shortly after Pearl Harbor it reverted to the War Department, and finally in 1946 it was returned to the Interior Department for park purposes. Rehabilitation of the old fort was begun in 1957 under the Mission 66 program.

Fort Washington is located on the Maryland side of the Potomac River. You can get there via Nichols Avenue and South Capitol Street S.E. to Indian Head Road (Maryland 210). Turn right five and a half miles below the district line and follow directional signs. There is also a small museum on the premises. Open daily from 8:00 A.M. until dusk. No admission fee.